W0055352

Fortschritte der Chemie organischer Naturstoffe

Progress in the Chemistry of Organic Natural Products

56

Founded by L. Zechmeister
Edited by W. Herz, H. Grisebach †, G.W. Kirby, and Ch. Tamm

Authors:
J. Asselineau, J. Kagan

Springer-Verlag
Wien New York 1991

Dr. W. HERZ, Professor of Chemistry, Department of Chemistry,
The Florida State University, Tallahassee, Florida, U.S.A.

G. W. KIRBY, Sc. D., Regius Professor of Chemistry, Chemistry Department,
The University, Glasgow, Scotland

Prof. Dr. CH. TAMM, Institut für Organische Chemie der Universität Basel,
Basel, Switzerland

Softcover reprint of the hardcover 1st edition 1991

Library of Congress Catalog Card Number AC 39-1015

Typesetting: Macmillan India Ltd., Bangalore-25

With 8 Figures

ISBN-13: 978-3-7091-9086-9 e-ISBN-13: 978-3-7091-9084-5
DOI: 10.1007/ 978-3-7091-9084-5

We deeply regret the death on March 29, 1990 of our fellow editor and friend Hans Grisebach. He was associated with our series since 1969 and contributed in no small measure to its continuing success. We shall miss his suggestions and advice.

W. Herz, Tallahassee
G. W. Kirby, Glasgow
Ch. Tamm, Basel

Contents

List of Contributors

ASSELINEAU, Dr. J., Centre de Recherche de Biochimie et Génétique Cellulaire, Université Paul Sabatier, 118 Route de Narbonne, F-31062 Toulouse, France.

KAGAN, Professor J., University of Illinois at Chicago, Department of Chemistry (M/C 111), P. O. Box 4348, Chicago, IL 60680, U.S.A.

Bacterial Lipids Containing Amino Acids or Peptides Linked by Amide Bonds

J. ASSELINEAU, Centre de Recherche de Biochimie et Génétique Cellulaires, Toulouse, France

With 6 Figures

Contents

I. Introduction

For a long time, amino acid or peptide containing lipids were considered as unusual components of bacteria. Isolation was mainly confined to "lipo-amino acids" which appeared to be transitory intermediates in cellular metabolism. Not much progress has been made on their metabolism.

More recently, thanks to improvements in fractionation and analytical techniques and because of the increase of interest in bacterial metabolites, particularly metabolites exhibiting antibiotic or enzyme inhibiting properties, many new peptidolipids have been discovered.

Whereas in the currently known cases peptides are always linked to fatty material by amide bonds, the amino acids in peptidolipids may be linked through amide or through ester bonds. The present review will be limited to compounds containing *amide bonds*. In the case of peptidolipids, an arbitrary limitation has been introduced: Only compounds having a fatty acid component of at least *ten carbon* atoms will be examined.

It is usual to distinguish between *peptidolipids* which exhibit lipid characteristics and are soluble in "lipid solvents" and *lipopeptides*, which exhibit polar characteristics and may be soluble in water. No such distinction has been made here because of the difficulty in applying these criteria to compounds that are insoluble in water as well as in chloroform. In fact, when derivatives of lipids and peptides are considered, no clear-cut separation into two classes of compounds becomes evident, but a progressive variation in the polarity of the molecules may be observed.

II. N-Acylamino Acids

As this review is devoted to derivatives of amino acids and peptides linked to fatty acids by amide bonds, lipids containing ester-bound amino acids such as ornithine esters of phosphatidylglycerol (*1*) or lysine esters of diglycerides (*2*) will not be considered.

The search for lipids containing amino acids in a lipidic bacterial extract makes it necessary to exclude the presence of free amino acids or peptides that can frequently exist as contaminants in lipid extracts. Water-soluble contaminants can be removed by washing a chloroform phase with methanol-water ("Folch wash") (*3*) or by passage of a chloroform solution of the crude extract through a column of Sephadex G-25 (*4*, *5*).

1. N-Acyl-L-serine

1.1. Serratamolide

A strain of *Serratia* resembling *S. indica* was isolated from coconuts and grown on a synthetic medium. An alkaline extract of the bacterial mass was prepared. After acidification, *serratamic acid* was obtained in the form of crystals m.p. 138 °C (dec.), $[\alpha]_D^{20} - 10°$ (ethanol) (*6*).

Acid hydrolysis of serratamic acid gave a nitrogen-free acid, $C_{10}H_{20}O_3$, and a water soluble component identical with L-serine. The acid, m.p. 47 °C, $[\alpha]_D^{20} - 19°$ (chloroform), was identified as 3(D)-hydroxydecanoic acid. As serratamic acid exhibits amide bands in its infrared spectrum (1649, 1632, 1549 cm^{-1}) and contains a primary hydroxyl group, structure (**1**), N-(3D-hydroxydecanoyl)-L-serine, was proposed (*6*) (Scheme 1). This structure was confirmed by synthesis from 3D-hydroxydecanoic acid and L-serine methyl ester (*7*).

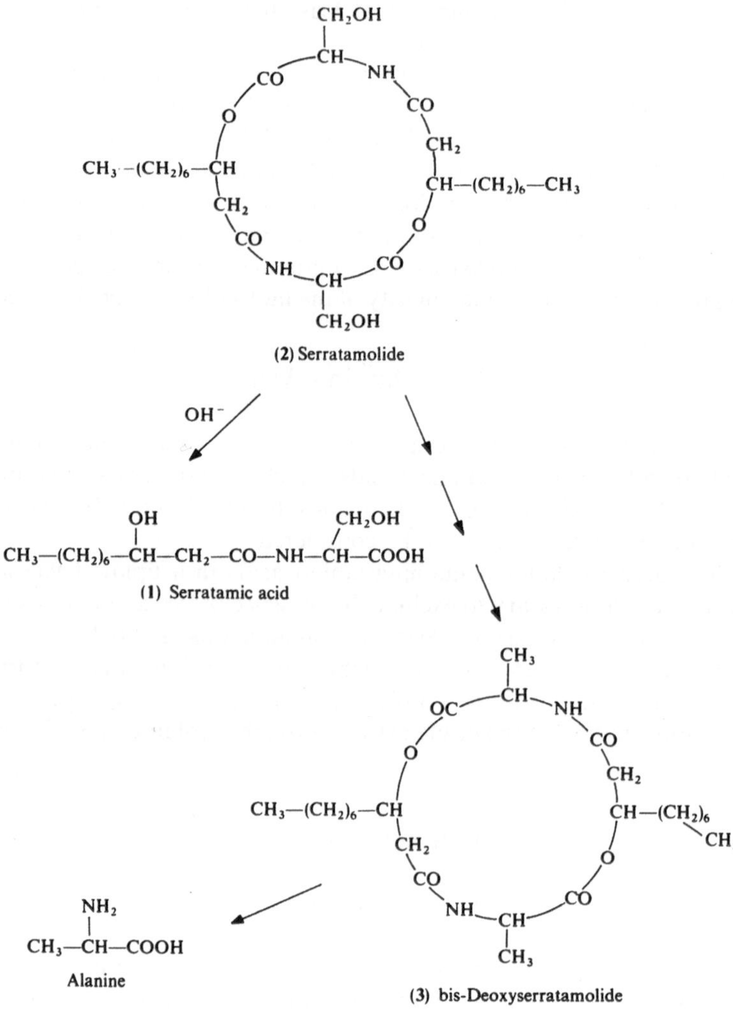

$$
\begin{array}{c}
\text{(2) Serratamolide: cyclic dimer, ring } -\text{CO}-\text{CH}(\text{CH}_2\text{OH})-\text{NH}-\text{CO}-\text{CH}_2-\text{CH}[(\text{CH}_2)_6\text{CH}_3]-\text{O}-\text{CO}-\text{CH}(\text{CH}_2\text{OH})-\text{NH}-\text{CO}-\text{CH}_2-\text{CH}[(\text{CH}_2)_6\text{CH}_3]-\text{O}-
\end{array}
$$

CH₂OH / CH / CO NH / O CO / CH₂ / CH₃–(CH₂)₆–CH CH–(CH₂)₆–CH₃ / CH₂ O / CO CO / NH CH / CH₂OH

(2) Serratamolide

OH^-

$$CH_3-(CH_2)_6-CH(OH)-CH_2-CO-NH-CH(CH_2OH)-COOH$$

(1) Serratamic acid

$$
\text{(3): ring } -\text{OC}-\text{CH}(\text{CH}_3)-\text{NH}-\text{CO}-\text{CH}_2-\text{CH}[(\text{CH}_2)_6\text{CH}_3]-\text{O}-\text{CO}-\text{CH}(\text{CH}_3)-\text{NH}-\text{CO}-\text{CH}_2-\text{CH}[(\text{CH}_2)_6\text{CH}_3]-\text{O}-
$$

(3) bis-Deoxyserratamolide

$$CH_3-CH(NH_2)-COOH$$

Alanine

Scheme 1

From an extract of *S. marcescens* with methylene chloride a neutral compound, M.W. 515, was isolated by fractional crystallization. Mild alkaline hydrolysis of this compound gave serratamic acid as the only product. The neutral compound, called *serratamolide*, of m.p. 159–160 °C and optical rotation $[\alpha]_D^{25}$ + 4.8° (ethanol), $C_{26}H_{46}N_2O_8$, was therefore a dimer of serratamic acid. Among three possible structures, structure **(2)** (Scheme 1) was suggested since hydrogenolysis of the dibromo derivative prepared by means of the dimesyl derivative gave a bis-deoxy derivative (**3**), the hydrolysis of which gave alanine as the only amino acid (*8, 9*).

This structure was confirmed by synthesis of the di-O,O′-acetyl derivative of serratamide (*10*) as shown in Scheme 2. The di-O,O′-acetyl-L-serine-L-serine anhydride (**4**) was acylated by means of 3D-benzyloxy-decanoyl chloride. Hydrogenolysis of the crude product gave a crystalline compound (30% yield) the properties of which were identical with those of di-O,O′-acetyl-serratamolide (**7**) described by WASSERMAN (*9*). Alkaline hydrolysis gave serratamic acid.

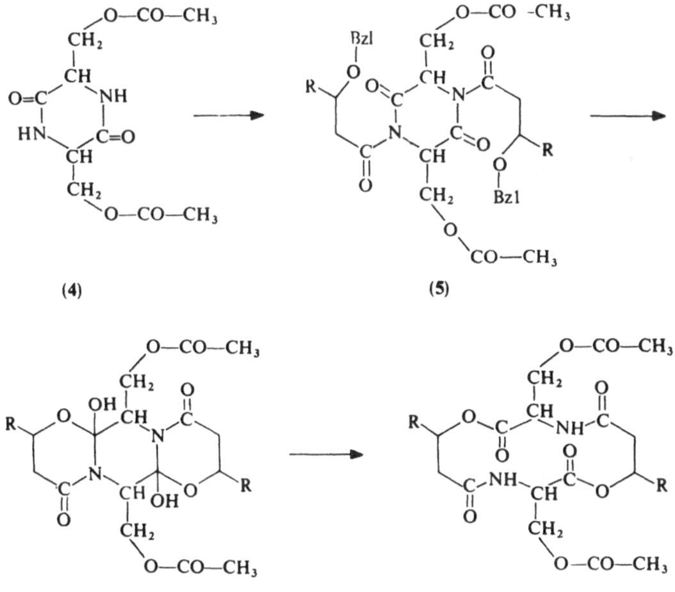

(4) **(5)**

(6) **(7)**

Bzl = C_6H_5—CH_2— R = CH_3—$(CH_2)_6$—

Scheme 2

A careful study of the acetone extracts of *S. marcescens* showed that serratamolide was the only lipidic bound form of serine and that serratamic acid was an artefact of alkaline extractions (*11*).

It was observed that the concentrations of serratamolide and prodigiosine, a tripyrrolic pigment produced by the same organism, responded similarly to changes in environmental and culture conditions: highly pigmented cells always had a large content of serratamolide. A common metabolic control might therefore exist (*12*).

1.2. Flavolipin

By thin layer chromatography of a lipid extract of a strain of *Flavobacterium meningosepticum* a serine-containing lipid was isolated. This lipid, for which the name flavolipin was proposed, contained 1 mole of ester-linked fatty acid (removed by mild saponification) and 1 mole of amide-linked fatty acid (requiring acid hydrolysis to make it free) for one mole of serine. These results and the information obtained from I.R. and mass spectra led the authors to propose structure (**8**) for flavolipin (*13*). No information was provided concerning the configuration of the serine molecule.

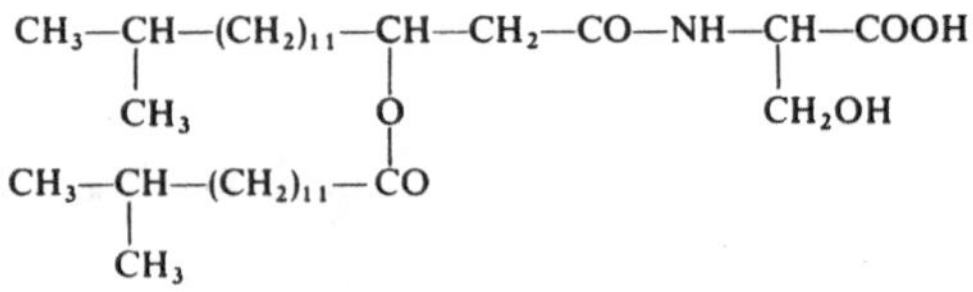

$$\begin{array}{l} CH_3{-}\underset{\displaystyle CH_3}{\underset{|}{CH}}{-}(CH_2)_{11}{-}\underset{\displaystyle \underset{\displaystyle CH_3{-}\underset{\displaystyle \underset{|}{CH_3}}{CH}{-}(CH_2)_{11}{-}CO}{|}}{\underset{|}{O}}}{\underset{|}{CH}}{-}CH_2{-}CO{-}NH{-}\underset{\displaystyle CH_2OH}{\underset{|}{CH}}{-}COOH \end{array}$$

(**8**) flavolipin

2. N-Acyl-L-ornithine

2.1. Ornithine-containing Lipid

The presence of ornithine in crude phospholipids of mycobacteria was demonstrated in 1955 (*14*). This amino acid was isolated after hydrolysis of the lipids and identified as L-ornithine by its chromatographic behaviour and the properties of derivatives (*14*).

Several years later, in the course of a study of mycobacterial lipids, it was shown that ornithine is not a component of the phospholipids, but is part of a new kind of compound devoid of phosphorus and containing fatty acids (*15*). Amide bands in its I.R. spectrum (1625 and 1550 cm^{-1}) made it likely that it was a derivative of N-acylornithine (*16*). Similar results were obtained in the course of a study of the lipids of *Rhodopseudomonas sphaeroides* (*17*).

The carboxyl group in the ornithine-containing lipid of *Rh. sphaeroides* seemed to be esterified, leading to an incorrect formula (*18*). In fact, the lipid behaved as a zwitterion. Acid hydrolysis gave a mixture of 3-hydroxy fatty acids, the major components of the mixture containing 16 and 18 carbon atoms, and non-hydroxylated fatty acids, the major components again containing 16 and 18 carbon atoms. Small peaks at m/z 688, 692 and 700 in the E.I.-mass spectrum considered to be the molecular peaks showed that hydroxylated fatty acids, non-hydroxylated fatty acids, and ornithine were present in the molecular ratio 1:1:1. Formula (**9**) was then proposed for the ornithine-containing lipid from *Rh. sphaeroides* (*19*).

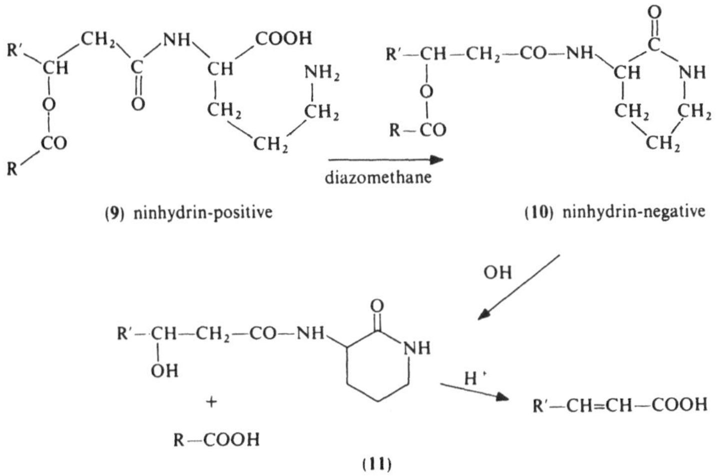

diazomethane

(**9**) ninhydrin-positive

(**10**) ninhydrin-negative

OH

$R'-CH(OH)-CH_2-CO-NH$ + R—COOH

(**11**)

H^+

R′—CH=CH—COOH

Scheme 3

The lipid was mainly located in the chromatophores and membrane material of *Rh. sphaeroides* (*20*). Similar results on its location were found in a reinvestigation of the lipids of subcellular fractions of *Rh. sphaeroides* (*21*); besides the ornithine-containing lipid a minor peptidolipid, designated as "aminolipid X", was observed. Hydrolysis of this lipid gave ornithine, an unidentified amine and fatty acids.

A large concentration of ornithine-containing lipid was found in a "basic membrane" material prepared from a mutant strain of *Rhodospirillum rubrum* (*22*). Studies on the structure of a similar lipid isolated from van Niels strain SI of *R. rubrum* (heterotrophically or autotrophically grown) were in agreement with structure (**9**) (*23*). In heterotrophically grown cells a second ornithine-containing lipid was found; it was stated that "the two lipids seem to be quite dissimilar in structure" (*23*).

Tables 1 and 2 show that ornithine-containing lipids are widely distributed in Gram negative bacteria. Many species of *Pseudomonas* have been studied and ornithine-containing lipids have been characterized in thirteen out of nineteen species (see Table 2). Such bacteria have two membrane layers, a cytoplasmic and an outer membrane. In the case of *Thiobacillus thiooxidans*, 1.9% of the polar lipids were ornithine-containing lipids (**12**). After disruption of the cells the two membranes

Table 1. *Distribution of Ornithine-containing Lipids (and a Lysine-containing Lipid) in Various Species of Gram Negative Bacteria*

	Ornithine-containing lipid with Structure of Type (**9**) major amide-linked acids	major ester-linked acids	Type (**13**)	References
Achromobacter	3-OH-18 3-OH-vaccenate	16:0		(*24*)
Bordetella pertussis *B. parapertussis,* *B. bronchiseptica*	3-OH-16	16:0		(*54–57*)
Brucella			**13***a*	(*31, 34, 38*)
Cytophaga johnsonae	3-OH-*i*17	-3-OH-*i*-15 -*i*-15		(*30, 58*)
Desulfovibrio gigas	3-OH-16	*i*-14, *n*-14, *a*-15, *i*-16, *n*-16, *n*-18		(*59*)
Erwinia aroideae	3-OH-16	*n*-16, *n*-18		(*49*)
Flavobacterium meningosepticum	3-OH-*i*17	-*i*-15 -2-OH-*i*-15		(*13, 50*)
Gluconobacter cerinus	3-OH-16	2-OH-16, 2-OH-vaccenate		(*46, 60, 61*)
Paracoccus denitrificans	3-OH-16	*n*-16		(*41, 62*)
Pseudomonas		see Table 2		
Rhodopseudomonas sphaeroides	3-OH-16 3-OH-18	16:0 18:0 18:1 19-cyclo		(*17–20*)
Rhodospirillum rubrum	3-OH-16	18:1 16:0 14:0		(*22, 23*)
Rhodomicrobium vannielii				(*39*)
Thiobacillus A2	3-OH-20:1 Δ^{13}	vaccenate		(*24*)
Thiobacillus ferrooxidans				(*25*)
Thiobacillus thiooxidans	3-OH-16	2-OH-lactobacil-late		(*26, 27, 47*)
Lysine-containing lipid:				
Agrobacterium tumefaciens	3-OH-16	16:0, vaccenate, lactobacillate		(*32*)

3-OH 16:0 = saturated fatty acid with 16 carbon atoms and a hydroxyl in position 3. cyclo = cyclopropane; *i* = iso; *n* = normal chain.

were separated by sucrose-density step gradient centrifugation. It was observed that about 80% of the ornithine-containing lipids were localized in the outer membrane (*27*). Similar results were obtained with *Cytophaga johnsonae* (*30*).

The two membranes of *Desulfovibrio gigas* were also separated by sucrose-density gradient centrifugation and analyzed for their amino-

Table 2. *Distribution of Ornithine-containing Lipids Among Various Species of Pseudomonas*

	Species containing ornithine-lipids			Species devoid of ornithine-lipids	References
Pseudomonas	Strain	Amide-linked acids	Ester-linked acids		
acidovorans				ATCC 15668	(*24*)
aeruginosa				ATCC 10145	(*63*)
aureofaciens	NCIB 9030				(*28*)
carboxydovorans				DSM 1227	(*24*)
cepacia	NCTC 10661	3-OH-16	– 16:0, 19-cyclo – 2-OH 18:1; 2-OH-19-cyclo		(*64*)
chlororaphis	NCTC 7357				(*28*)
delafieldii				ATCC 17505	(*24*)
denitrificans	NCIB 8376				(*28*)
facilis				ATCC 17695	(*24*)
fluorescens	NCTC 10038				(*28, 29, 63*)
fragi	NCIB 8542				(*28*)
mucidolens	NCTC 8068				(*28*)
ovalis	NCTC 912				(*28*)
putida	NCIB 9034			ATCC 12633	(*24, 28*)
*rubescens**	NCIB 8768	3-OH-16	*i*-15		(*65*)
stutzeri	NCIB 9040				(*28, 63*)
syncyanea	NCTC 9943				(*28*)
synxantha	NCIB 8178				(*28*)
taetrolens	NCIB 9396				(*28*)

* This strain is now designated as *Shewanella putrefaciens*.

Abbreviations of culture collections: ATCC: American Type Culture Collection; NCIB: National Collection of Industrial Bacteria, England; NCTC: National Collection of Type Cultures, England; DSM: Deutsche Sammlung von Mikroorganismen.

Table 3. *Occurrence of Ornithine-containing Lipids in Gram Positive Bacteria*

Bacterial species	Ornithine-containing lipids: Amide-linked fatty acids	Ornithine-containing lipids: Ester linked fatty acids	References
*Mycobacterium tuberculosis**	3-OH-18	tuberculostearate	(*38*)
bovis (BCG)	"	"	(*31*, *38*)
*marinum***	"	"	(*38*)
sp. 1217			(*16*)
Streptomyces			
660-15		2-OH-*i*-15, 2-OH-*i*-16	(*40*)
aureoverticillus	3-OH-*a*-17, 3-OH-*i*-17 3-OH-*i*-16	*a*-15, *i*-16, *a*-17, *a*-15	(*40*)
globisporus	3-OH-*i*-16, 3-OH-*i*-17	*a*-15	(*40*)
olivaceus	3-OH-*i*-16, 3-OH-*i*-17	*a*-15	(*40*)
sioyaensis	3-OH-*i*-16, 3-OH-*i*-17	2-OH-*i*-15	(*36*, *66*, *67*, *68*)
toyokaensis	3-OH-*i*-16, 3-OH-*i*-17	2-OH-*i*-15	(*37*)

* No ornithine-containing lipid could be detected in strain H37 Ra.

** No ornithine-containing lipid was found in *M. phlei* (*24*, *38*); another kind of lipid, devoid of phosphorus and containing ester-bound lysine was identified (*2*).

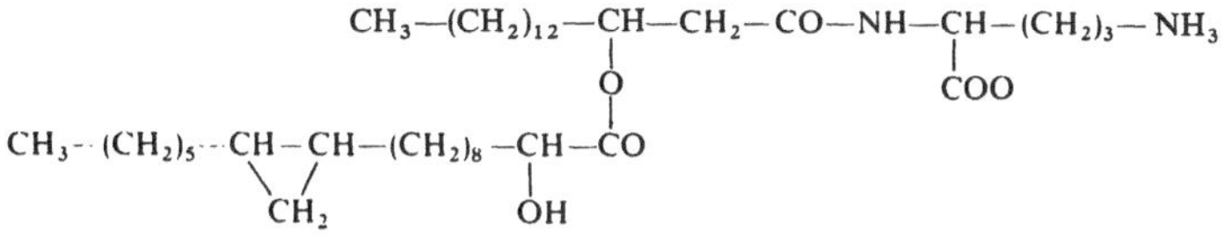

$$
\begin{array}{l}
CH_3{-}(CH_2)_{12}{-}CH{-}CH_2{-}CO{-}NH{-}CH{-}(CH_2)_3{-}NH_3 \\
\qquad\qquad\qquad\quad |\qquad\qquad\qquad\qquad\quad | \\
\qquad\qquad\qquad\quad O\qquad\qquad\qquad\qquad COO \\
\qquad\qquad\qquad\quad | \\
CH_3{-}(CH_2)_5{-}CH{-}CH{-}(CH_2)_8{-}CH{-}CO \\
\qquad\qquad\quad \backslash\,/ \qquad\qquad\quad | \\
\qquad\qquad\quad CH_2 \qquad\qquad\quad OH
\end{array}
$$

(12)

lipid, phosphatidylglycerol and phosphatidylethanolamine content. The level of ornithine-containing lipid per milligram of protein was 385 nanomoles in the cytoplasmic membrane and 350 nanomoles in the outer membrane. In this bacterium, the distribution was about the same in the two membrane layers (*59*).

While ornithine-containing lipids are widely distributed in Gram negative bacteria, so far they have only been found in a small number of Gram positive species (Table 3), all of them belonging to the Actinomycetales order.

Most of these ornithine-containing lipids have a structure of type (**9**) (Scheme 3), where a 3-hydroxy fatty acid (usually 3-hydroxyhexadecanoic acid) is linked by an amide bond to the α-amino group of ornithine. A second acid, either a 2-hydroxy fatty acid or a non-hydroxylated fatty acid, esterifies the hydroxyl group of the first acid. Therefore saponification frees the ester-linked acid. Because of the structure of the ornithine molecule, the ornithine residue is easily converted into a piperidone (**10**), a cyclization catalyzed by acid or alkali. Cyclization proceeds even when attempts are made to methylate the carboxyl group with diazomethane (*24*, *31*) as the ester is more easily attacked by the ω-amino group than the free carboxyl. When *Cytophaga johnsonae* was grown in liquid medium, the ester-linked fatty acid was a non-hydroxylated one, but when surface grown cells were used, 3-hydroxylated ester-linked acids were formed (*30*). So far *C. johnsonae* is the only organism in which a 3-hydroxy fatty acid is found ester-linked; usually the ester-linked fatty acid has its hydroxyl group in position 2 (Table 1).

An almost identical lipid in which ornithine is completely replaced by lysine has been isolated from *Agrobacterium tumefaciens*. Here again the α-amino group of lysine was linked by an amide bond to the usual 3-acyloxyhexadecanoic acid (*32*). In siolipin, an ornithine-containing lipid produced by *Streptomyces sioyaensis* and *S. toyocaensis* (Table 3), lysine was found together with ornithine in the ratio 9:91 (*33*).

It is sometimes very difficult to isolate pure ornithine-containing lipids from accompanying lipids, in particular from phosphatidylethanolamine, perhaps because of ionic linkages. Hence they are often contaminated by other lipid components. This phenomenon probably affords an explanation for the fact that the first analyses performed on the ornithine-containing lipids of *Mycobacterium bovis* (B.C.G.) and *Brucella melitensis* consistently showed the presence of α-diols (1,2-ethanediol and 1,2-propanediol respectively) in the hydrolysate. For this reason, structures (**13***a* and **13***b*) were proposed originally (*31*). This type of structure was also proposed for the ornithine-containing lipids isolated from *Brucella species* (*34*, *35*) and for siolipin from *S. sioyaensis* and *S. toyocaensis* (*36*, *37*). However, recent investigations of the ornithine-containing lipids of several species of *Mycobacterium* and of *B. melitensis* by mass spectrometry established that these lipids are devoid of any α-diol moiety and have in fact the usual structure (**9**) (*38*). It is most probable that *all* the ornithine-containing lipids with an amide bond have the same type of structure.

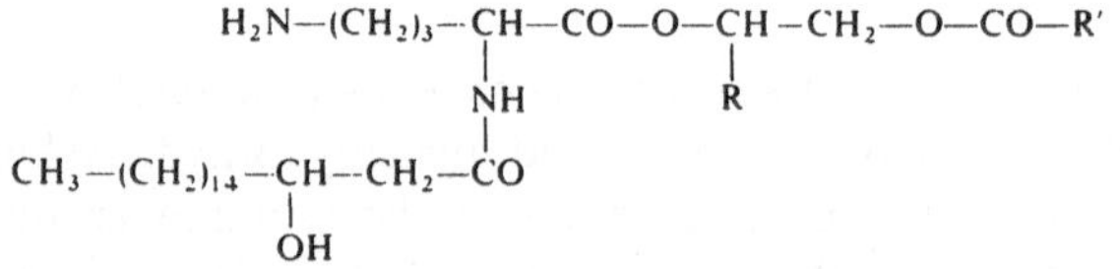

$$
\begin{array}{l}
H_2N{-}(CH_2)_3{-}\underset{\substack{|\\NH\\|}}{CH}{-}CO{-}O{-}\underset{\substack{|\\R}}{CH}{-}CH_2{-}O{-}CO{-}R' \\
CH_3{-}(CH_2)_{14}{-}\underset{\substack{|\\OH}}{CH}{-}CH_2{-}CO
\end{array}
$$

(13) *a*: $R = CH_3$
$R'{-}COOH$ = lactobacillic acid
b: $R = H$
$R'{-}COOH$ = tuberculostearic acid

Six fractions containing ornithine were observed in the lipids isolated from different species of *Rhodospirillaceae*. One of them contained phosphorus and was probably an ester of ornithine and phosphatidylglycerol. Little information was provided for the other fractions except for the R_f values in thin-layer chromatography (*39*).

Purification and fractionation of ornithine-containing lipid homologues has been performed by gas-chromatography of their TMS derivatives. Because of the low volatility of these compounds, a short column has to be used (*13*).

The general structure (**9**) ascribed to the ornithine-containing lipids with two hydrophobic acyl chains and a zwitterionic group shows the amphiphilic character of the molecule. This structure is quite similar to that of phosphatidylethanolamine (Fig. 1).

When the culture medium of *Pseudomonas fluorescens* was made phosphate-limited, phosphatidylethanolamine disappeared and the ornithine-containing lipids increased (*29*). Similar observations were made in the case of the lipids of *Streptomyces* strains (*40*). These observations on the similarity of structure and the easy interchange

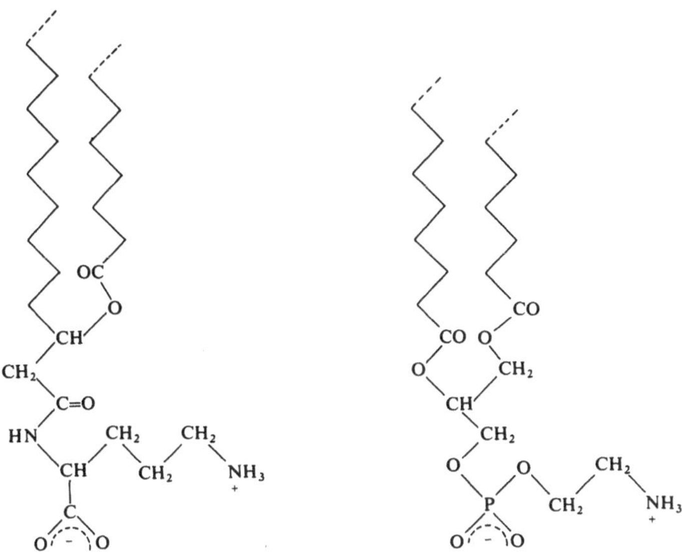

Fig. 1. Conformations of ornithine-containing lipid (left) and phosphatidylethanolamine (right), at the interface air: water

between phosphatidylethanolamine and ornithine-containing lipids are in agreement with the fact that their location in the membrane of bacterial cells is the same.

Cells of *Paracoccus denitrificans* grown in a complex medium deficient in divalent cations exhibited a higher ratio of ornithine-containing lipid to phospholipids than that observed in cells grown in the same medium supplemented with Mg^{2+} and Ca^{2+}. It has been suggested that the zwitterionic ornithine-containing lipid was less dependent than acidic phospholipids on divalent cations for their incorporation in the outer membrane (*41*).

Membrane vesicles prepared from *Thiobacillus ferrooxidans* contained three enzymes of the iron oxidation system. Delipidation of these vesicles by aqueous acetone decreased the enzymatic activities. They were restored by incubation of the lipid-depleted vesicles with a dispersion of ornithine-containing lipid together with coenzyme Q_8. A possible role of the ornithine-containing lipid in the iron oxidation system has been suggested (*25*).

Little information is available on the metabolism of lipoamino acids in general and of N-acylornithine in particular. From a cell-free homogenate of *Gluconobacter capsulatus*, a particulate enzyme was obtained which was able to catalyze the transfer of an acyl group from acyl-coenzyme A to the hydroxyl group of N^{α}-(3-hydroxyhexadecanoyl)-L-ornithine, thus giving an ornithine-containing lipid. The acceptor was prepared by reaction of 3-hydroxyhexadecanoylsuccinimide and N^{δ}-carbobenzyloxy-L-ornithine. The experiment was performed with palmitoyl-coenzyme A: no information was given on the specificity of the enzyme preparation as regards the acyl substrate (*42*). So far nothing is known of the formation of N^{α}-(hydroxyacyl)-L-ornithine.

However enzymes able to hydrolyse amides of amino acids and long-chain fatty acids (C_{12}–C_{18}) were detected in various bacterial species, particularly in *Gluconobacter* and *Pseudomonas* (*43*). These enzymes, under suitable conditions, would have the capacity of synthesizing such N-acyl-amino acids. The enzymes of *Pseudomonas diminuta* were more thoroughly studied. Two enzyme preparations were obtained: long-chain acylamino acylase I (M.W. about 300,000 by gel filtration) was specific for N-acyl derivatives of L-glutamic acid (*43*), while long-chain acylamino acylase II (M.W. 220,000) was able to hydrolyse N-acyl derivatives of glutamic acid and various other amino acids (*44*).

2.2. Cerilipin

During a study of the lipids of *Gluconobacter cerinus* it was observed that an ornithine-containing lipid was labelled with sulphur-35 when [^{35}S]-sulphate was added to the culture medium.

By thin-layer chromatography this lipid was isolated and named cerilipin. On hydrolysis, it gave ornithine and a sulphur-containing component identified as taurine. By alkaline transmethylation, cerilipin was deacylated and the resulting methyl ester was found to be identical with methyl 2-hydroxyhexadecanoate (with a small amount of methyl 2-hydroxylactobacillate). Acid hydrolysis of the deacylated lipid gave a fatty acid identified as 3-hydroxyhexadecanoic acid as well as ornithine and taurine, in the molecular ratio 1:1:1. Treatment of cerilipin with dinitrofluorobenzene gave N^{δ}-ornithine as the only DNP derivative. From these data, structure (**14**) was proposed for cerilipin (*45*).

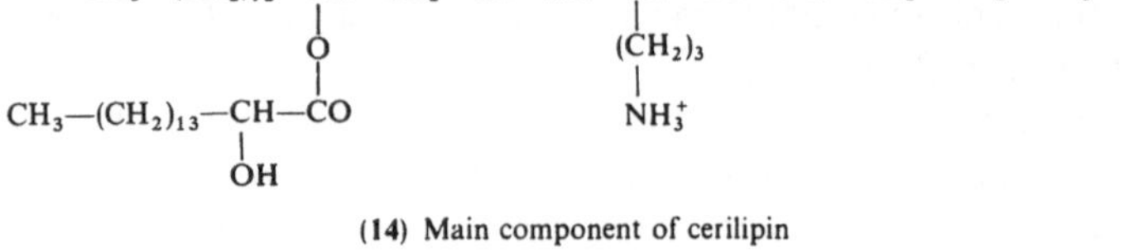

$$
\begin{array}{l}
CH_3\text{—}(CH_2)_{12}\text{—}CH\text{—}CH_2\text{—}CO\text{—}NH\text{—}CH\text{—}CO\text{—}NH\text{—}CH_2\text{—}CH_2\text{—}SO_3^- \\
\qquad\qquad\qquad\quad | \qquad\qquad\qquad\qquad\quad | \\
\qquad\qquad\qquad\quad O \qquad\qquad\qquad\quad (CH_2)_3 \\
\qquad\qquad\qquad\quad | \qquad\qquad\qquad\qquad\quad | \\
CH_3\text{—}(CH_2)_{13}\text{—}CH\text{—}CO \qquad\qquad\quad NH_3^+ \\
\qquad\qquad\qquad | \\
\qquad\qquad\qquad OH
\end{array}
$$

(**14**) Main component of cerilipin

A particulate fraction prepared from *Gluconobacter cerinus* cells was able to catalyze the condensation of ornithine-containing lipid and taurine to produce cerilipin, when the medium was supplemented with ATP and Mn^{2+} (*46*).

3. Mass Spectrometry

During the first analyses of ornithine-containing lipids, the use of mass spectrometry was limited to the characterization of hydrolysis products because of the state of the technique at the time. GORCHEIN (*19*)

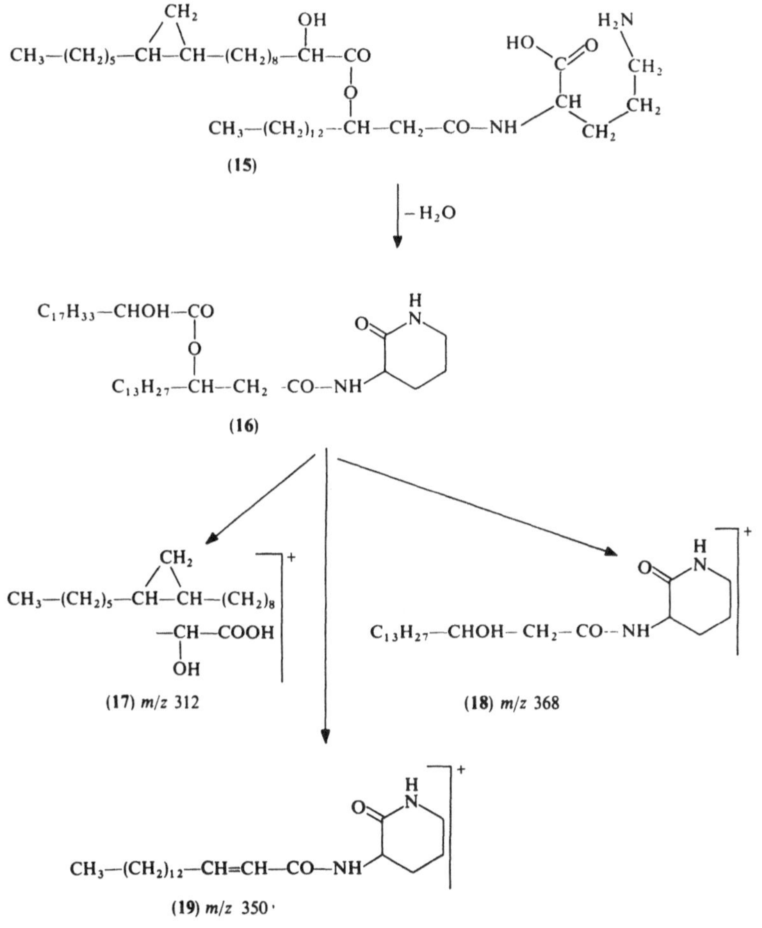

$CH_3-(CH_2)_5-CH(CH_2)CH-(CH_2)_8-CH(OH)-CO-O-CH(CH_2CH_2 ...)$

$CH_3-(CH_2)_{12}-CH-CH_2-CO-NH-CH(COOH)-CH_2-CH_2-CH_2-NH_2$

(**15**)

$-H_2O$

$C_{17}H_{33}-CHOH-CO-O-CH(C_{13}H_{27})-CH_2-CO-NH-$(piperidinone)

(**16**)

$CH_3-(CH_2)_5-CH(CH_2)CH-(CH_2)_8-CH(OH)-COOH]^+$

(**17**) *m/z* 312

$C_{13}H_{27}-CHOH-CH_2-CO-NH-$(piperidinone)$]^+$

(**18**) *m/z* 368

$CH_3-(CH_2)_{12}-CH{=}CH-CO-NH-$(piperidinone)$]^+$

(**19**) *m/z* 350

Scheme 4

was probably the first to observe peaks corresponding to the ionized whole molecule in an electron-impact mass spectrum of free ornithine-containing lipid (*m/z* 688, 692, 700).

A careful study of the ions formed from the free ornithine-containing lipid of *Th. thiooxidans* (**15**) in E.I. mass spectrometry has been performed (*47*). The lipid, in underivatized form, was introduced into the mass spectrometer source by gentle pyrolysis from the direct inlet probe. Elimination of a molecule of water gave a 2-piperidone derivative (**16**) sufficiently volatile to give mass spectra. By operating at 15 eV the high-mass region was quite informative, exhibiting a prominent peak at *m/z* 350 corresponding to ion (**19**), and minor peaks at *m/z* 312 (**17**) and 368 (**18**) (Scheme 4). Spectra obtained by thermolysis of this lipid in a chemical ionization source using methane or isobutane exhibited a base peak at *m/z* 351 corresponding to the protonated ion (**19**) (*48*).

The protonated molecular ions of ornithine-containing lipids isolated from *Erwinia aroideae* were observed by chemical ionization mass spectrometry with ammonia of the derivatives prepared by treatment with dimethylformamide dimethylacetal (**20**) (*49*).

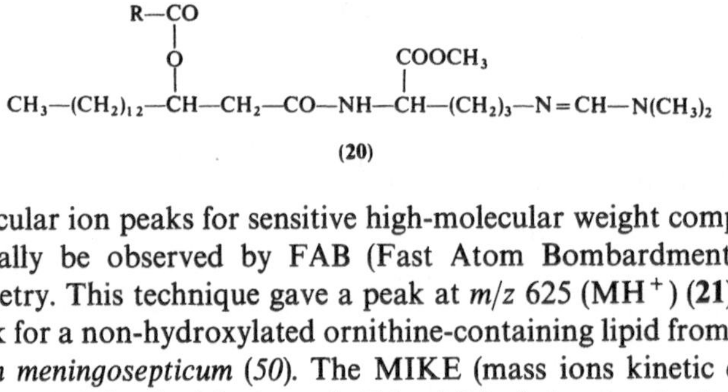

$$\begin{array}{l} \qquad\qquad\quad R\text{—}CO \\ \qquad\qquad\qquad\quad | \\ \qquad\qquad\qquad\quad O \qquad\qquad\qquad\qquad\qquad COOCH_3 \\ \qquad\qquad\qquad\quad | \qquad\qquad\qquad\qquad\qquad\quad | \\ CH_3\text{—}(CH_2)_{12}\text{—}CH\text{—}CH_2\text{—}CO\text{—}NH\text{—}CH\text{—}(CH_2)_3\text{—}N{=}CH\text{—}N(CH_3)_2 \end{array}$$

(20)

Molecular ion peaks for sensitive high-molecular weight compounds may usually be observed by FAB (Fast Atom Bombardment) mass spectrometry. This technique gave a peak at *m/z* 625 (MH^+) (**21**) as the base peak for a non-hydroxylated ornithine-containing lipid from *Flavobacterium meningosepticum* (*50*). The MIKE (mass ions kinetic energy) mass spectrum (Fig. 2) of the ion *m/z* 625 exhibited three peaks due to the ion fragments at *m/z* 401 (**22**), 383 (**23**) and 365 (**24**). As shown in Scheme 5, formation of these ions is due to elimination of isopentadecanoic acid and demonstrates the presence of 3-hydroxyisopentadecanoic acid amide-linked to ornithine. The hydroxylated ornithine-containing lipid gave a FAB mass spectrum with a base peak at *m/z* 641 (MH^+) and a peak at *m/z* 383 corresponding to the elimination of a *hydroxy*-isopentadecanoic acid. The hydroxy fatty acids were identified by EI-mass spectrometry of their methyl esters (*50*).

Similar results were obtained by CID-MIKE (collision induced dissociation) mass spectrometry of an ornithine-containing lipid (*51*). Homologues were detected by FAB mass spectrometry and the ions were distinguished by their CID spectra (*52*).

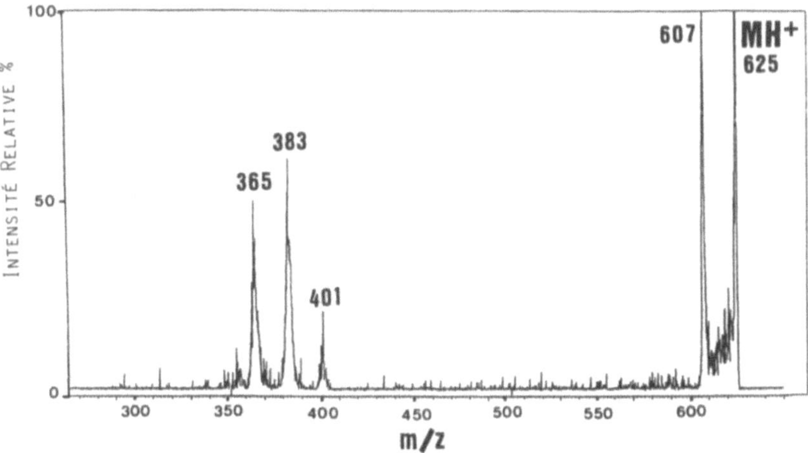

Fig. 2. MIKE mass spectrum of the ion *m/z* 625 generated by Fast Atom Bombardment (FAB) from the ornithine-containing lipid of *Flavobacterium meningosepticum* (*50*)

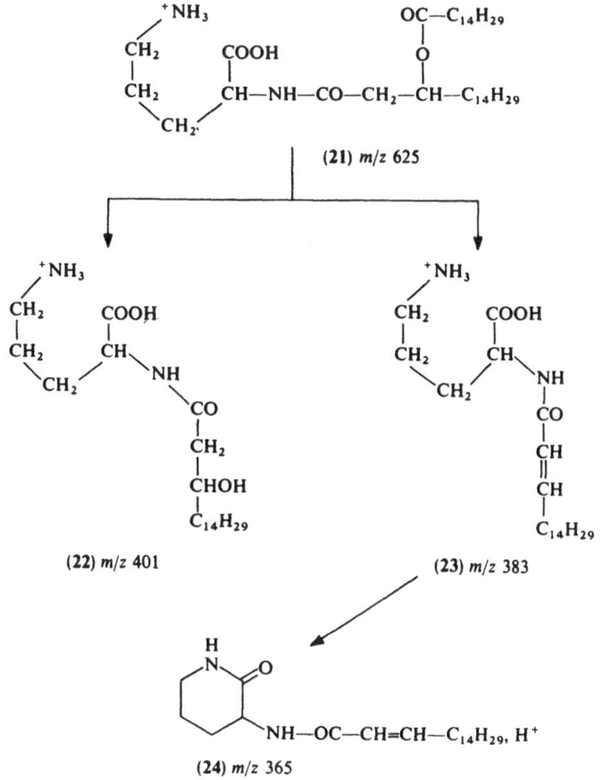

$^{+}NH_3$–CH_2–CH_2–CH_2–CH(COOH)–NH–CO–CH_2–CH(O–OC–$C_{14}H_{29}$)–$C_{14}H_{29}$

(**21**) *m/z* 625

$^{+}NH_3$–CH_2–CH_2–CH_2–CH(COOH)–NH–CO–CH_2–CHOH–$C_{14}H_{29}$

(**22**) *m/z* 401

$^{+}NH_3$–CH_2–CH_2–CH_2–CH(COOH)–NH–CO–CH=CH–$C_{14}H_{29}$

(**23**) *m/z* 383

H, N, O (piperidin-2-one ring) –NH–OC–CH=CH–$C_{14}H_{29}$, H^+

(**24**) *m/z* 365

Scheme 5

The free lipids of *Fl. meningosepticum* were particularly rich in glycine, as well as in serine and ornithine (*50*). So far no lipophilic derivative of glycine has been identified.

4. Biological Properties

Siolipin, the hydroxylated ornithine-containing lipid of *Streptomyces sioyaensis*, was found to specifically inhibit the growth of a strain of *Bacillus subtilis* when a synthetic culture medium was used. No inhibition was observed when the bacteria were grown on complex medium, as L-histidine and to a lesser degree L-cysteine were able to antagonize the effect of siolipin (*53*).

Ornithine-containing lipids exhibited two kinds of activities on erythrocytes, hemolysis and hemagglutination. Hemolytic activity, measured on rabbit erythrocytes, was detected at concentrations lower than 1 μg per milliliter (*53*). Human erythrocytes (type A and B) were agglutinated at pH 6.0 by ornithine-containing lipid at concentrations as low as 1 to 2 μg per milliliter. Only weak agglutination was observed when erythrocytes of chicken, horse, sheep or guinea pig were used. When the ornithine-containing lipid was O-deacylated, the hemagglutinating property was lost, but hemolytic activity remained (*54*, *55*).

Flavolipin was found to be more active than ornithine-containing lipids in the hemagglutination test: the activity was observed at concentrations lower than 1 μg per milliliter (*13*).

Ornithine-containing lipid was shown to accelerate the coagulation of blood in a thrombine-fibrinogen system, by using a concentration of 24 μg per ml (*53*).

In conclusion, few biological properties have so far been detected for ornithine-containing lipids in spite of their amphiphilic character.

III. N-Acylpeptides

N-acylpeptides may show a large diversity of structures and properties due in part to the varying proportions in the size of the fatty acid and the peptide moiety. Most often the N-acylpeptides do not contain a C-terminal amino acid bearing a free carboxyl group. The terminal carboxyl is either methylated or engaged in a diketopiperazine ring in the case of linear peptide moieties or esterified by the hydroxy group of a fatty acid or a serine/threonine residue, or amidified by the amino group of a β-amino fatty acid, giving rise to lactone or lactam rings.

Esterification or amidification of the terminal carboxyl group affects only slightly the general properties of the N-acylpeptides when other free carboxyl groups provided by aspartic or glutamic acid residues are present. These modifications contribute to the hydrophobic character of the whole molecule. Lactone or lactam rings are frequently encountered, a feature that plays a part in maintaining the peculiar conformation necessary to the biological role of the molecules. For instance, simple opening of lactone rings causes disappearance of the antibiotic properties.

D-Amino acids are frequently encountered in these bacterial constituents. The formation of D-*allo*threonine can be explained by the action of racemases on the α-carbon atom of L-threonine (*69*). Similarly, D-*allo*isoleucine can be produced from L-isoleucine (*70*) (see for example peptidolipin NA or cerexin A). However, in two cases, L-*allo*threonine and L-*allo*isoleucine have been found; no information is available on their biosynthesis. Unusual amino acids have also been observed in some of these peptidolipids, for instance pipecolic acid (amphomycin), L-*threo*-β-hydroxyglutamic and L-*threo*-β-methylglutamic acids (neopeptines), or D- or L- 2,4-diaminobutyric acid (octapeptines).

1. N-Acylpeptides with a Linear Peptide Moiety

1.1. N-Acyldipeptides: Majusculamides A and B

A peptidolipid fraction was found in a chloroform extract of *Lyngbya majusculata* Gomont (Cyanobacteria) collected in shallow marine water near the island of Oahu. These peptidolipids were isolated both from dermatitis and non-dermatitis producing varieties, but were not produced by *L. gracilis*. Final purification of the crude product by HPLC gave two isomeric substances designated as majusculamides A and B, $C_{28}H_{45}N_3O_5$ (*71*).

Majusculamide A had m.p. 96–97°C, $[\alpha]_D^{26}$ + 19° (ethanol) and majusculamide B, m.p. 102–103°C, $[\alpha]_D^{26}$ + 15° (ethanol). Their molecular weight was determined by FD mass spectrometry (M^+ 503). EI mass spectrometry gave peaks at *m/z* 486 and 487 corresponding to the loss of NH_3 and NH_2, suggesting that these compounds were primary amides in agreement with their neutrality.

By heating majusculamide B in dimethylsulfoxide at 140°C for 15 hours, 2-methyl-3-oxodecanoic amide (**26**) and the diketopiperazine formed from N,O-dimethyl-tyrosyl-N-methylvaline (**27**) were obtained. Acid hydrolysis of the majusculamides under mild conditions gave

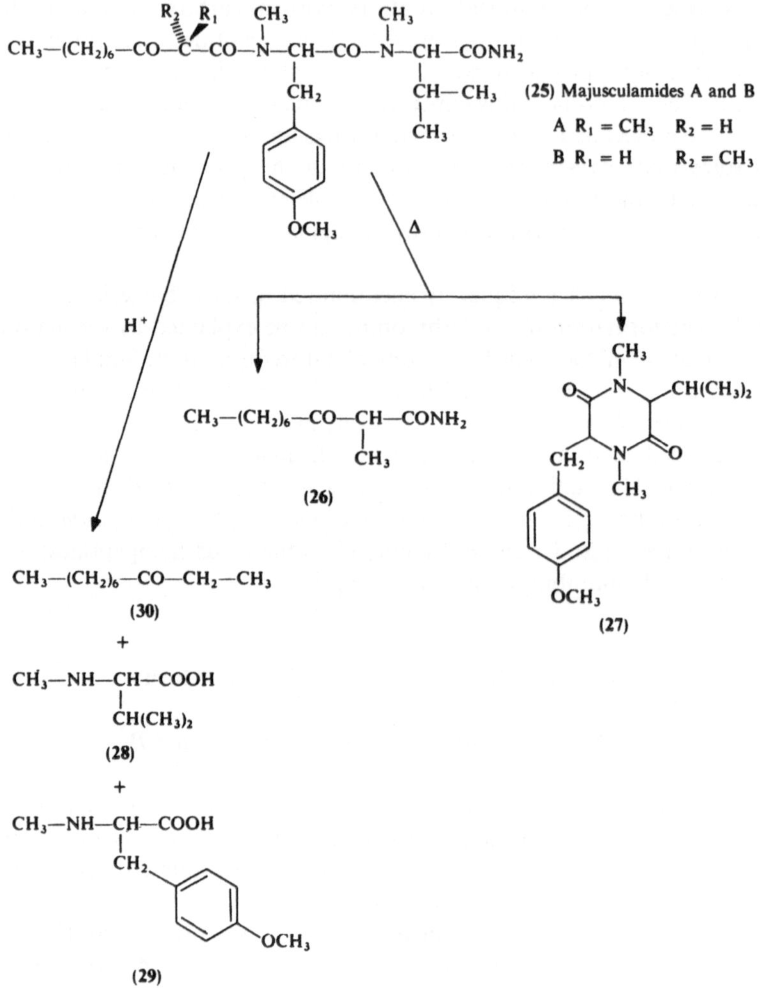

CH_3—$(CH_2)_6$—CO—C(R_1)(R_2)—CO—N(CH_3)—CH(CH_2—C_6H_4—OCH_3)—CO—N(CH_3)—CH(CH(CH_3)—CH_3)—$CONH_2$

(25) Majusculamides A and B

A $R_1 = CH_3$ $R_2 = H$

B $R_1 = H$ $R_2 = CH_3$

H⁺ / Δ

CH_3—$(CH_2)_6$—CO—CH(CH_3)—$CONH_2$

(26)

(27) [diketopiperazine: N-CH_3, $CH(CH_3)_2$, O, N-CH_3, O, CH_2—C_6H_4—OCH_3]

CH_3—$(CH_2)_6$—CO—CH_2—CH_3

(30)

\+

CH_3—NH—CH($CH(CH_3)_2$)—COOH

(28)

\+

CH_3—NH—CH(CH_2—C_6H_4—OCH_3)—COOH

(29)

Scheme 6

L-(+)-N-methylvaline (**28**), D(−)-N,O-dimethyltyrosine (**29**) and 3-decanone (**30**), the latter arising by decarboxylation of an intermediate β-keto acid (Scheme 6).

From these data, structure (**25**) was proposed for the majusculamides and was confirmed by X-ray crystallographic studies on the B isomer. Majusculamides A and B are isomers differing only in the stereochemistry of the asymmetric center in position 2 of the keto fatty acid (*71*).

1.2. N-Acyltetrapeptides

Acidic peptidolipids were isolated from the paraffin-oxidizing bacterium *Rhodococcus erythropolis* (previously known as *Mycobacterium paraffinicum*). Chromatography separated the peptidolipids of low polarity from the more polar ones. Further chromatography separated the low polarity group into three components whose structures were studied by chemical degradation and spectroscopy of the natural products and of model compounds. In two of these peptidolipids the hydroxyl group of an L-threonine residue, the N-terminal amino acid of the peptide moiety, esterified the carboxyl group of either normal chain fatty acids or of C_{28}–C_{42} mycolic acids (**31**) (*72, 73*). It is rare to find a mycolic acid esterifying the hydroxyl group of an amino acid residue. In the third member of this group of peptidolipids, the carboxyl of the C-terminal amino acid leucine esterified the hydroxyl of the threonine residue, forming a lactone ring (*74*).

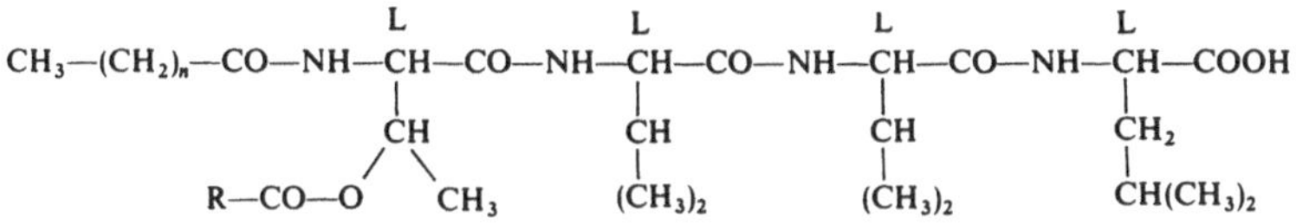

```
                            L               L               L               L
CH3—(CH2)n—CO—NH—CH—CO—NH—CH—CO—NH—CH—CO—NH—CH—COOH
                  |               |               |               |
                  CH              CH              CH              CH2
                 /  \             |               |               |
        R—CO—O      CH3        (CH3)2          (CH3)2          CH(CH3)2
```

(**31**) $n = 18$–26

R—COOH = normal chain fatty acids C_{20}–C_{28}
or mycolic acids C_{28}–C_{42}

Hydrolysis of the polar peptidolipids which consisted of four components gave glycine, L-leucine, D-*allo*isoleucine, L-threonine, L-serine, L-homoserine and D-alanine in the molar ratio 3:3:2:2:2:1:1. Two of these compounds contained glucose bound to the hydroxyl group of serine or threonine and thus were glucosides of peptidolipids; all of them contained mycolic acids (*74*). Their complete structures have not yet been established.

1.3. N-Acylpentapeptides

From a lipid extract of *Mycobacterium paratuberculosis* a peptidolipid fraction was isolated thanks to its insolubility in ether. Chemical analysis showed the presence of the amino acids D-phenylalanine, L-phenylalanine, L-leucine, L-isoleucine and the presence of a methyl ester group at the C-terminal end of the peptide moiety. Presence of an N-acyltetrapeptide was suggested (*75*). However, mass spectrometric investigation showed that the substance was in fact an N-acylpentapeptide and led to the complete sequence of amino acids (**32**) (*76*) (see the mass spectrum, Fig. 4, page 55). The peptide moiety contained a Leu(Ile)-Leu(Ile) dipeptide, very difficult to hydrolyse, which explained the discrepancies in the results obtained by chemical analysis.

$$CH_3{-}(CH_2)_n{-}CO{-}NH{-}\underset{\substack{|\\ CH_2\\ |\\ C_6H_5}}{\overset{D}{CH}}{-}CO{-}NH{-}\underset{\substack{|\\ CH_2\\ |\\ CH(CH_3)_2}}{\overset{L}{CH}}{-}CO{-}NH{-}\underset{\substack{|\\ CH_2\\ |\\ CH(CH_3)_2}}{\overset{L}{CH}}{-}CO{-}$$

$${-}NH{-}\underset{\substack{|\\ CH_2\\ |\\ C_6H_5}}{\overset{L}{CH}}{-}CO{-}NH{-}\underset{\substack{|\\ CH_3}}{\overset{L}{CH}}{-}COOCH_3$$

(**32**) $n = 18, 16$

The N-terminal amino acid was acylated by a fatty acid molecule. Acid hydrolysis gave a mixture of saturated fatty acids, all with an even number of carbon atoms (C_{16}-C_{24}; major members C_{20} and C_{18}). Partial hydrolysis gave N-acyl-D-phenylalanine, so the location of the D-amino acid residue became known (*77*). The hydrophobic character of this peptidolipid is to be noted.

The naturally occurring N-acylpentapeptide methyl ester had a m.p. 138–139°C, $[\alpha]_D - 87°$ (chloroform). By chromatography on alumina it was partially saponified, leading to isolation of the free acid of m.p. 171–174°C, $[\alpha]_D - 30°$ (chloroform). Methylation of the acid by diazomethane gave a methyl ester $[\alpha]_D - 86°$. Such a great difference in optical rotation of an acid and its methyl ester is unusual; for instance, the methyl ester of N-octadecanoyl-D-phenylalanine has $[\alpha]_D - 60°$ and the free acid $[\alpha]_D - 54°$.

From the results obtained by physical studies, it was suggested that the acid form of the N-acyl peptide was a dimer able to form an α-helix, whereas the methyl ester was a monomer with a different conformation (*78*). It would be interesting to study this pair of compounds using modern NMR spectroscopic methods. As the reversible transformation acid $\rightleftharpoons$ ester is easily performed in a living cell, the resulting change of conformation might play a role in the permeability of the cell envelope.

1.4. N-Acyloctapeptide: Stenothricin

Aerobic fermentation of a common medium by *Streptomyces griseus* produces an antibiotic called *stenothricin* which was isolated by ion-exchange chromatography and purified by gel filtration. This antibiotic has a narrow spectrum and inhibits cell wall biosynthesis (*79*).

Hydrolysis permitted isolation and identification of eight amino acids: N-methylglycine (sarcosine), three L-amino acids (valine, serine, lysine), two D-amino acids (cysteic acid, serine) as well as D-*allo*threonine and L-2,3-diaminopropionic acid. Their configuration was determined by capillary gas chromatography on an optically active phase. No fatty acid was isolated by hydrolysis; only methylketones were obtained, showing that the acyl part was derived from β-keto-acids with iso, anteiso and normal chains of length C_{14}–C_{17}.

Partial hydrolysis gave a mixture of di-, tri- and tetrapeptides which was fractionated and studied by gas chromatography/mass spectrometry in the form of trifluoroacetyl methyl ester derivatives. The complete structure is shown in formula (**33**) (*80*).

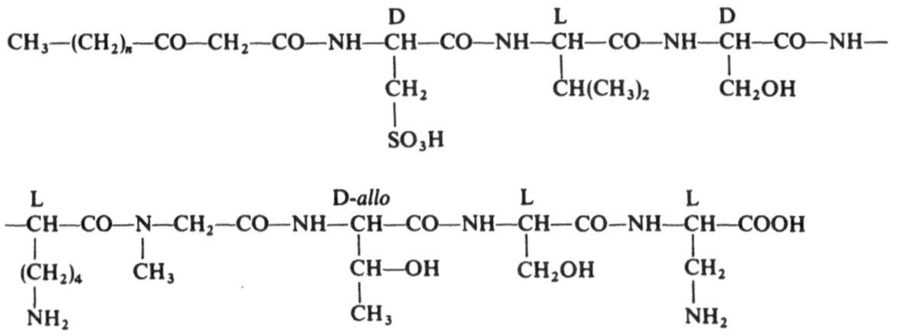

$$CH_3\text{—}(CH_2)_n\text{—}CO\text{—}CH_2\text{—}CO\text{—}NH\text{—}\overset{D}{CH}(CH_2\text{—}SO_3H)\text{—}CO\text{—}NH\text{—}\overset{L}{CH}(CH(CH_3)_2)\text{—}CO\text{—}NH\text{—}\overset{D}{CH}(CH_2OH)\text{—}CO\text{—}NH\text{—}$$

$$\text{—}\overset{L}{CH}((CH_2)_4\text{—}NH_2)\text{—}CO\text{—}N(CH_3)\text{—}CH_2\text{—}CO\text{—}NH\text{—}\overset{D\text{-}allo}{CH}(CH\text{—}OH\text{—}CH_3)\text{—}CO\text{—}NH\text{—}\overset{L}{CH}(CH_2OH)\text{—}CO\text{—}NH\text{—}\overset{L}{CH}(CH_2\text{—}NH_2)\text{—}COOH$$

(**33**) n = 9–13 (iso, anteiso or normal chain) Stenothricin

1.5. N-Acylnonapeptide: Fortuitin

A peptidolipid called fortuitin was isolated from *M. fortuitum*. After purification, it had a m.p. of 199–202°C and an optical rotation of $[\alpha]_D - 72°$ (chloroform). Acid hydrolysis furnished approximately equimolecular quantities of eicosanoic and docosanoic acid and amino acids. The amino acid composition was found to be Val_3, Thr_2, Ala_1, Pro_1. The carboxyl group of the C-terminal amino acid of the oligopeptide moiety was esterified by methanol. A preliminary structure was reported (*81*).

In fact fortuitin contained a sequence of branched-chain amino acids, in particular two molecules of N-methylleucine, making complete hydrolysis very difficult as in the case of the acylpeptide isolated from *M. paratuberculosis*. Mass spectrometry gave the exact amino acid sequence and showed that the hydroxyl groups of the two threonine residues were esterified by acetic acid. The complete structure of fortuitin is given in formula (**34**) (*82*). It should be noted that fortuitin, like the two acylpeptides of mycobacterial origin discussed earlier, is quite hydrophobic; the only polar groups within the molecule, the hydroxyls of the threonine residues, are esterified. The L configuration was demonstrated for all amino acids except for N-methylleucine where no conclusion could be reached.

```
                         (1)        CH3 (2)             (3)              (4)       CH3
                                     |                                              |
CH3—(CH2)n—CO—NH—CH—CO—N—CH—CO—NH—CH—CO—NH—CH—CO—N—
                  |          |           |            |
               CH(CH3)2     CH2       CH(CH3)2     CH(CH3)2
                             |
                          CH(CH3)2

 (5)          (6)            (7)           (8)          (9)  COOCH3
—CH—CO—NH—CH—CO—NH—CH—CO—NH—CH—CO—N—<pyrrolidine ring>
  |           |              |             |
 CH2          CH             CH            CH3
  |          /  \           /  \
CH(CH3)2    O   CH3         O   CH3
            |               |
           CO—CH3          CO—CH3
```

(**34**) $n = 18, 20$ Fortuitin

^{1}H-NMR spectroscopy at 400 MHz indicated that in pyridine fortuitin exhibits a hairpin conformation, with the bend occurring between Val-3 and Val-4. This folded conformation seems to be in equilibrium with extended forms (*83*). The conformation and self-association of fortuitin is strongly dependent on the polarity of the solvent. These observations must be kept in mind when studying the role of fortuitin inside the cell envelope; however nothing is known of its precise location.

1.6. N-Acyldecapeptides: Cerexins

Bacillus cereus produces a peptidolipid fraction exhibiting antibiotic activity against Gram positive bacteria which has been called cerexin. Cerexin is a complex mixture of peptidolipids of similar structures (Table 4). The structure of the main product, cerexin A, was established first (*84, 85*). It behaved like an amphoteric compound and had an optical rotation of $[\alpha]_D + 19.5°$ (dimethylformamide). Short acid hydrolysis gave a fatty acid the methyl ester of which exhibited a single peak on gas chromatography, while more prolonged hydrolysis gave four acids. Mass spectrometry of the single peak ester showed that it had a hydroxyl group in position 3; in particular a peak was observed at *m/z* 103 (ion $[CHOH–CH_2–COOCH_3]^+$) characteristic of β-hydroxy fatty acid methyl esters. The prolonged hydrolysis resulted in dehydration and lactonization, thus giving rise to supplementary peaks on gas chromatography of the methyl esters.

Complete hydrolysis of cerexin A gave seven different amino acids whose configurations were studied by enzymatic methods. In the case of tryptophan and *allo*isoleucine the assignments were based on ORD curves. γ-Hydroxylysine was identified as the L-*threo* isomer (*84*). As only one acidic group was found by alkalimetric titration and as the C-terminal amino acid (*allo*isoleucine) was liberated by treatment with N-bromosuccinimide, the aspartic acid residues were probably present as asparagine; this was confirmed by reduction of the carboxamide groups to methylene amino groups (*86*). Some difficulties were encountered with the Asn-Asn sequence (*85*).

The peptide moiety of cerexin A was isolated by enzymatic deacylation using polymyxine acylase of *Pseudomonas* sp. M 6-3 (*87*). This enzyme had a rather broad specificity; about 20% of cerexin A could be deacylated.

Table 4. *Composition of the Peptidolipids of the Cerexin Group* (*from 88*)

$R–CHOH–CH_2–CO$-D.Asn-D.Val- X_3 -L.Asn-D.Asn-X_6-D.*allo*Thr- X_8 -D.Trp-D.*allo*Ile-OH

Cerexin	Fatty acid β-OH	X_3 D	X_6 L	X_8
A	i-C_{11}	Val	γHyl	L. Ser
B_1	i-C_{10}	Phe	γHyl	Gly
B_2	n-C_{10}	Phe	γHyl	Gly
B_3	a-C_{11}	Phe	γHyl	Gly
B_4	i-C_{11}	Phe	γHyl	Gly
C	i-C_{11}	Val	Lys	L. Ser
D_1	i-C_{10}	Phe	Lys	Gly
D_2	n-C_{10}	Phe	Lys	Gly
D_3	a-C_{11}	Phe	Lys	Gly
D_4	i-C_{11}	Phe	Lys	Gly

i = iso *a* = anteiso
γHyl = γ-hydroxylysine

Concentrated hydrochloric acid cleaved the peptide chain after the γ-hydroxylysine residue. Edman degradation of these various peptides permitted determination of the complete sequence of amino acids and led to proposal of structure (**35**) for cerexin A (*85*). Further work (*88*) showed that the cerexin fraction could be resolved into ten components, the structures of which were established (Table 4).

$$\begin{array}{l} \qquad\qquad\qquad\quad OH \\ \qquad\qquad\qquad\quad | \\ (CH_3)_2CH—(CH_2)_8—CH—CH_2—CO—\text{D. Asn-D. Val-D. Val-L. Asn-D. Asn-L. } \gamma\text{HyL} \\ \qquad\qquad\qquad\qquad\qquad\qquad\qquad\qquad\qquad\qquad\qquad\qquad\qquad\qquad\qquad\qquad \downarrow \\ \qquad\qquad\qquad\qquad HO—\text{D. } allo\text{Ile} \longleftarrow \text{D. Trp} \longleftarrow \text{L. Ser} \longleftarrow \text{D. } allo\text{Thr} \end{array}$$

(**35**) Cerexin A

1.7. N-Acyltridecapeptides: Tridecapeptins

A discussion of this group of lipopeptides produced by *B. polymyxa* is introduced here because of the great similarities to the cerexins with respect to bacterial sources and general structural features. However the tridecapeptins which have a high basic amino acid content due to the presence of diaminobutyric acid have hydrophilic character and are soluble in water. They exhibit antibiotic properties against Gram positive and Gram negative bacteria and can be connected with the polymyxins.

The methods used for the structure determination of cerexins were extended to the study of the tridecapeptins (*89, 90*). High performance

Table 5. *Composition of the Lipopeptides of the Tridecapeptin Group* (*from 90*)

R—CO- X_1 -D.DAB-Gly-D.Ser-D.Trp-L.Ser-L.DAB-D.DAB- X_9 -L.Glu-D.Val- X_{12} -X_{13} -OH

Tridecapeptin	Fatty acid	X_1	X_9 L	X_{12} D	X_{13} L
Aα	β-OH *a*-C_9	D. Val	Phe	*allo*Ile	Ala
Aβ	β-OH *a*-C_9	D. Val	Phe	Val	Ala
Bα	*a*-C_9	Gly	Ile	*allo*Ile	Ser
Bβ	*a*-C_9	Gly	Ile	Val	Ser
Bγ	*a*-C_9	Gly	Val	*allo*Ile	Ser
Bδ	*a*-C_9	Gly	Val	Val	Ser
Cα1	β-OH *a*-C_9	D. Val	Phe	Val	Ser
Cα2	β-OH *i*-C_{10}	D. Val	Phe	Val	Ser
Cβ	β-OH *a*-C_{11}	D. Val	Phe	*allo*Ile	Ser

liquid chromatography resolved the tridecapeptin fraction into nine components. Their structures are given in Table 5 (*88*).

1.8. N-Acylundecapeptide: Amphomycin

A family of closely structurally related antibiotics is produced by several strains of *Streptomyces* (*e.g. Str. violaceus, griseus-spiralis, viridochromogenes, parvulus, pseudogriseolus*). The most studied member of this family is amphomycin which was shown to be identical with glumamycin and aspartocin (*91*).

After purification by counter-current distribution this peptidolipid was obtained as an amorphous powder of m.p. 230°C (dec.) and optical rotation $[\alpha]_D + 9.5°$ (ethanol). Acid hydrolysis gave a mixture of fatty acids, the major components of which were identical with (+)-anteisotridec-3-enoic and isododec-3-enoic acids. The amino acids consisted of L-aspartic acid (3 moles), glycine (2 moles), L-*threo*-β-methylaspartic acid, L-proline, L-valine, D-pipecolic acid, L-*threo*-2,3-diaminobutyric acid (DAB_t) and D-*erythro*-2,3-diaminobutyric acid (DAB_c) (1 mole each).

By boiling for a short time in dilute acetic acid specific splitting of the molecule occurred after a residue of aspartic acid. N-Acylaspartic acid was isolated as well as a hexapeptide (**36**), the structure of which was established by mass spectrometry of the acetylated, permethylated or perdeuteriomethylated derivatives.

Gly-DAB_c-Val-Pro-DAB_t-Pip (**36**)
(ring linking DAB_t and Pip)

MeAsp-Asp-Gly-Asp-Gly-DAB_c-Val-Pro-DAB_t--Pip (37)
(ring linking DAB_t and Pip)

Spontaneous hydrolysis of amphomycin in water solution over about two months was observed at room temperature, probably because of the acidity provided by the free carboxyl groups. A decapeptide (**37**) and N-acylaspartic acid were thus obtained. Treatment of the decapeptide with nitrous acid destroyed the methylaspartic acid residue and transformed DAB_c into threonine. Dansylation showed that methylaspartic acid was the only N-terminal amino acid; the fact that no derivative was obtained from pipecolic acid or DAB_t supported the presence of a diketopiperazine. The possibility that the diketopiperazine could have been formed under mild conditions of degradation was ruled out by experiments

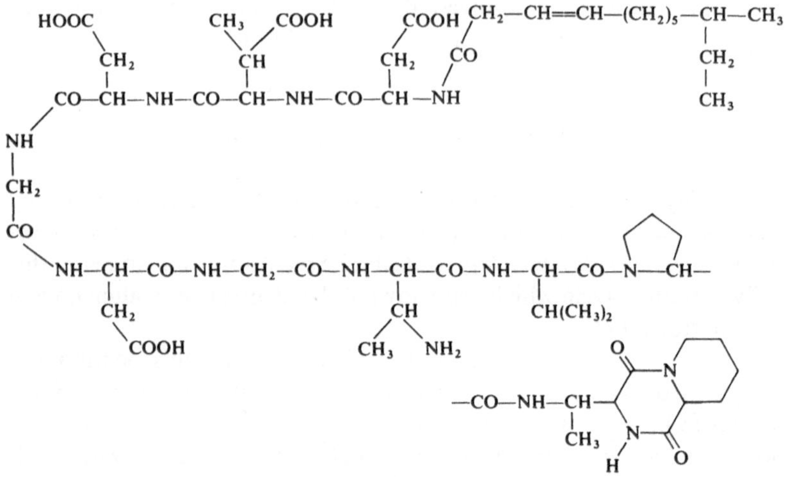

```
HOOC             CH3  COOH         COOH CH2—CH=CH—(CH2)5—CH—CH3
    \              \ /             /   /                  |
     CH2            CH          CH2  CO                  CH2
     |              |            |   /                    |
  CO—CH—NH—CO—CH—NH—CO—CH —NH                            CH3
 /
NH
|
CH2
|
CO
  \
   NH—CH—CO—NH—CH2—CO—NH—CH—CO—NH—CH—CO—N—CH—
      |                  |          |     (pyrrolidine)
      CH2                CH         CH(CH3)2
        \               /  \
         COOH        CH3    NH2

                            —CO—NH—CH—(diketopiperazine–piperidine ring)
                                   |
                                   CH3
```

(38) Amphomycin

on model peptides. Finally, structure (**38**) was proposed for amphomycin (*91*).

Several other peptidolipids differing from amphomycin only in the structure of the fatty acid constituent are listed in Table 6.

Table 6. *Antibiotics Structurally Related to Amphomycin*

Antibiotic	Bacterial strain	Fatty acid component
Tsushimycin	*Str. pseudogriseolus*	*anteiso*-$C_{15:1}$, Δ^3 *iso*-$C_{14:1}$, Δ^3, *n*-C_{14} *iso*-$C_{15:1}$, Δ^3
Zaomycin or Parvulin	*Str. zaomyceticus*	*iso*-$C_{12:1}$, iso-$C_{13:1}$ *anteiso*-$C_{13:1}$, Δ^3
Laspartomycin	*Str. viridochromogenes*	*iso*-$C_{15:1}$, *trans* Δ^2

2. N-Acylpeptides Containing a Hydroxy Fatty Acid Involved in a Lactone Ring

2.1. N-Acylpentapeptide: Globomycin

This peptidolipid has been found in four species of Actinomycetales, *Streptomyces halstedii*, *Streptoverticillium cinnamoneum*, *Streptomyces neohygroscopicus* var. *globomyceticus* and *Streptomyces hagronensis*. It has antibiotic activity against Gram negative bacteria by inhibiting cell wall synthesis.

Globomycin was obtained as colourless crystals of m.p. 115 °C and optical rotation $[\alpha]_D \sim 0°$ (chloroform). Its structure was established by means of the degradations depicted in Scheme 7. Four amino acids were obtained by acid hydrolysis, glycine, serine, *allo*threonine and *allo*-isoleucine. *Allo*threonine and *allo*isoleucine were identified by comparison with authentic samples. It was also observed that in the mass spectra of the trimethylsilyl derivatives of threonine and *allo*threonine, the relative intensities of the ions *m/z* 218 and 219 were significantly different and could be used for identification.

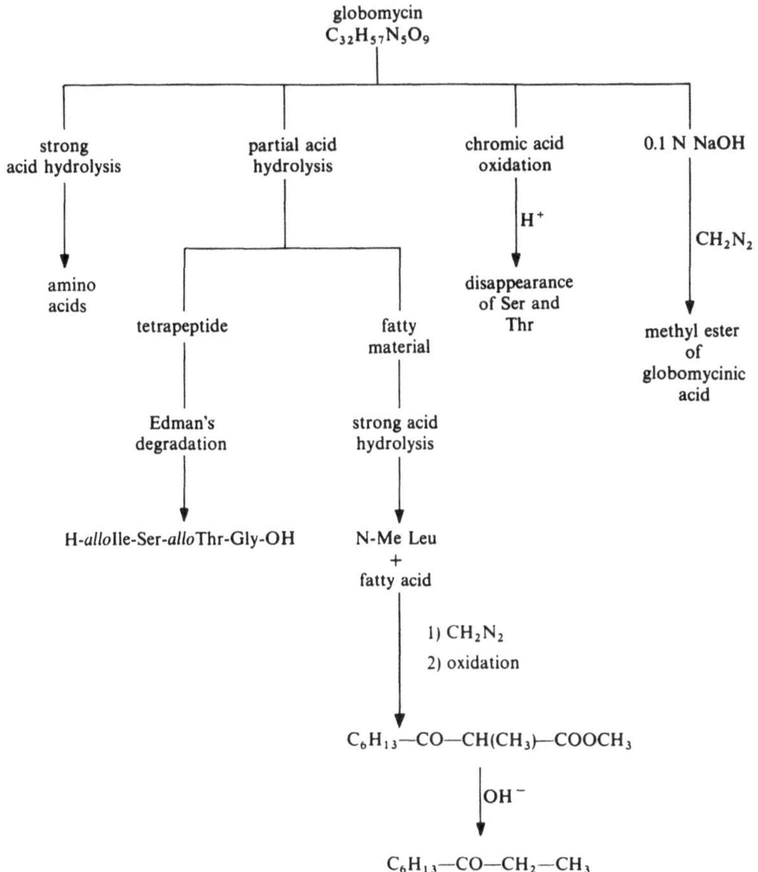

Scheme 7. Chemical degradation of globomycin (from *92*)

As the amide bond with the fatty acid was more resistant to acid hydrolysis than the peptide bonds, it was possible to isolate N-acyl-N-methyl-leucine by acid hydrolysis which permitted identification of the fifth amino acid. Its isolation was in agreement with the presence of N-methyl protons as indicated by a signal at 3.2 ppm in the ^{1}H-NMR spectrum of globomycin. The fatty acid was found to be 2-methyl-3-hydroxynonanoic acid, which furnished 3-nonanone by oxidation (see Scheme 7).

The FAB mass spectrum of globomycin gave a molecular weight of 655 ($C_{32}H_{57}N_5O_9$). The presence of a lactone ring was indicated by the transformation of globomycin to globomycinic acid on mild treatment with alkali. Chromic acid oxidation of globomycin caused disappearance of the serine and threonine units but left the hydroxyl group of the fatty acid residue unchanged, thus leading to the conclusion that the hydroxyl group of the hydroxy fatty acid was engaged in the lactone ring.

All these results supported structure (**39**) for globomycin (*92*). The amino acids were isolated and from their optical rotation their configuration was determined. It should be noted that the occurrence of L-*allo*threonine and L-*allo*isoleucine in nature is rather rare. No information was provided regarding the stereochemistry of 2-methyl-3-hydroxynonanoic acid in globomycin.

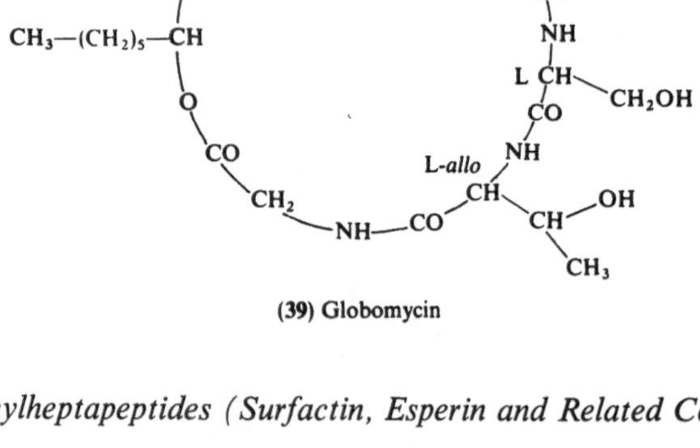

(**39**) Globomycin

2.2. *N-Acylheptapeptides (Surfactin, Esperin and Related Compounds)*

By screening more than 100 strains of Gram positive bacteria for the presence of compounds with surfactant activity, a "protoplast-bursting factor" or surfactin was isolated from a strain of *Bacillus subtilis* (*93*). The pure product had m.p. 138–140 °C, $[\alpha]_D^{27}$ + 40° (chloroform) and −39° (methanol) (*94*).

Hydrolysis of surfactin gave one mole of 3-hydroxy-isopentadecanoic acid and the following amino acids, L-aspartic acid, L-glutamic acid, L-valine, L-leucine and D-leucine, in the molar ratio 1:1:1:2:2. By partial hydrolysis ten peptides were obtained which were separated and analyzed, thus providing the amino acid sequence. This sequence was checked by Edman degradation which showed that aspartic and glutamic acids were bound through their α-carboxyl groups. The presence of a lactone ring was demonstrated on mild treatment with alkali.

Surfactin had three carboxyl groups able to participate in the lactone ring, the ω-carboxyl groups of aspartic and glutamic acids and the carboxyl group of the C-terminal L-leucine. In order to locate the lactone ring, reduction with $LiBH_4$ was performed on intact surfactin, on surfactin after methylation with diazomethane and on its product of saponification. Each product was hydrolyzed and the amino acid composition was determined. In the first case, one of the four leucine residues disappeared and in the second case, glutamic and aspartic acids and also one leucine disappeared. In the third case, no change of amino acid composition was observed. It was concluded that the carboxyl group of the C-terminal amino acid (L-leucine) esterified the hydroxyl group of the fatty acid, thus leading to structure (**40**) (*96*). The sequence of amino acids

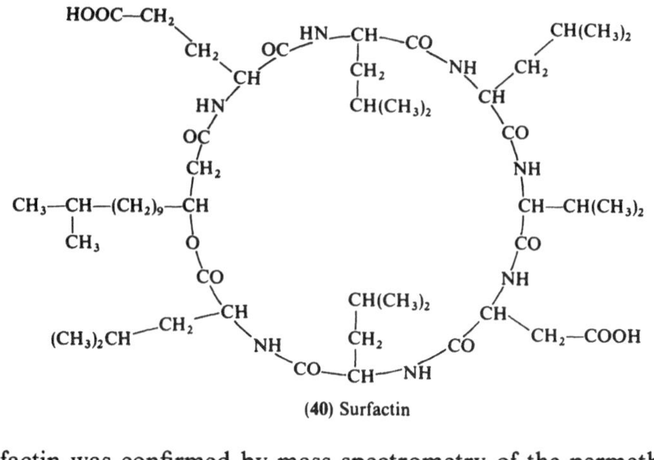

(**40**) Surfactin

in surfactin was confirmed by mass spectrometry of the permethylated derivative of its saponification product with an opened lactone ring (*97*).

The optimal rate of surfactin synthesis occurs at the end of the exponential growth phase. A study of the incorporation of labelled amino acids into surfactin showed that activation of the amino acids by ATP was a preliminary step. An enzyme able to catalyze ATP-P_i exchange in the presence of the specific amino acids was isolated (*98*). The production of surfactin seemed to depend on the carbon–nitrogen balance of the culture medium and was inversely correlated with biomass production (*99*).

An incompletely characterized peptidolipid with the same amino acid composition as surfactin has been isolated from the Marburg strain of *Bacillus subtilis*, and appeared to be located in the cell membrane. It was synthesized continuously during the growth of the bacteria and its content increased in phosphate-limited cultures (*100*).

In the course of screening for inhibitors of cyclic adenosine-3′,5′-monophosphate phosphodiesterase, a strain of *B. subtilis* was found to produce a series of such inhibitors. These compounds with a m.p. of 136–139 °C and an optical rotation of $[\alpha]_D^{22}$ + 37° to 38.5° (chloroform), − 33° to − 38° (methanol), were lipopeptides with the same peptide moiety as surfactin. Their fatty acid components were 3-hydroxy-11-methyldodecanoic acid, 3-hydroxy-10-methyldodecanoic acid, 3-hydroxytetradecanoic acid (the major component), 3-hydroxy-13-methyltetradecanoic acid (the same acid as that of surfactin) and 3-hydroxy-12-methyltetradecanoic acid (*101*, *102*). The location of the lactone ring was established by the same method as that used for surfactin (*96*). Therefore these inhibitors had structures quite similar to, and for one of them identical with, that of surfactin (**40**).

From *B. mesentericus* a peptidolipid exhibiting antibiotic properties was isolated and called *esperin*. It had m.p. 238 °C and $[\alpha]_D$ − 24° (methanol). Mild saponification opened a lactone ring, thus giving esperinic acid of m.p. 195 °C, $[\alpha]_D$ + 12.5° (methanol). A tentative structure (**41**) was proposed (*103*). However, because of differences in the properties of esperinic acid and a synthetic sample of the peptidolipid (**41**) the proposed structure was found to be incorrect (*104*).

R-CHOH-CH_2-CO-L. Glu-L. Asp-L. Val-L. Leu-D. Leu-OH (**41**)

R-CHOH-CH_2-CO-Glu-Leu-Leu-Val-Asp-Leu-Leu-OH

(**42**) Esperinic acid R = $C_{12}H_{25}$ (45%)
$C_{11}H_{23}$ (35%)
$C_{10}H_{21}$ (20%)

Esperinic acid was studied by mass spectrometry and the revised structure (**42**) was established (*105*), in which 30% of the C-terminal L-leucine was replaced by L-valine. In order to locate the lactone ring esperin was treated with hydrazine followed by Curtius rearrangement and hydrolysis. This treatment transformed esterified aspartic acid into α,β-diaminopropionic acid. The revised structure of esperine is thus (**43**) (*105*). The only difference between surfactin (**40**) and esperin (**43**) is the size of the lactone ring. It would have been interesting to have a direct comparison between the saponification product of surfactin and esperinic acid, which should be identical, as well as a comparison of the biological properties of surfactin and esperin.

HOOC—$(CH_2)_2$—CH—CO—NH
CH_2—$CH(CH_3)_2$
NH CH
CO CO
CH_2 NH $CH(CH_3)_2$
CH_2
$C_{12}H_{25}$—CH CH
O CO
CO NH
CH_2 CH $CH(CH_3)_2$
CH—NH—CO
HOOC—CH—NH—OC—CH—NH—OC
CH_2 CH_2
$(CH_3)_2CH$ $CH(CH_3)_2$

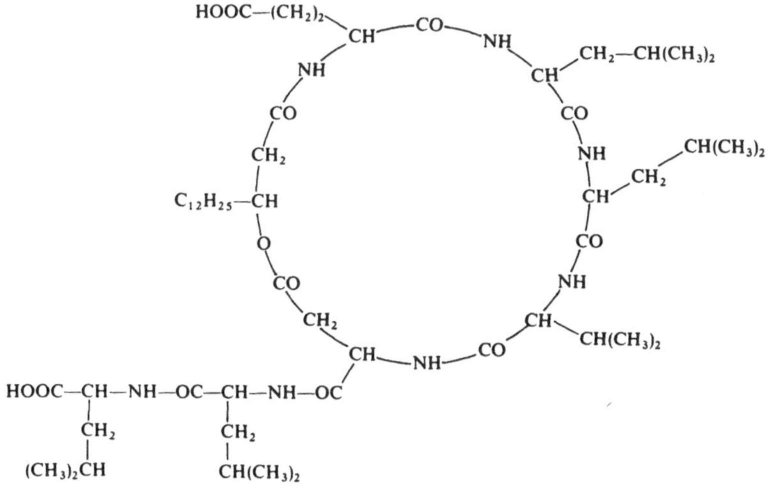

(**43**) Esperin

2.3. N-Acylheptapeptides: Peptidolipin NA

A peptidolipid fraction was isolated from *Nocardia asteroides* ATCC 9969. The main component of the mixture, obtained by counter-current distribution, was called *peptidolipin NA*. After purification, it had m.p. 232–233 °C, $[\alpha]_D$ + 42° (chloroform) (*106*).

The amino acids obtained after acid hydrolysis were identified as L-alanine, D-alanine, L-valine, D-*allo*isoleucine, L-threonine and L-proline in the molar ratio 1:1:1:1:2:1. Partial acid hydrolysis gave a hexapeptide the structure of which was established by chemical methods as L-Thr-L-Val-D-Ala-L-Pro-D-*allo*Ile-L-Ala (*107*). Strong acid hydrolysis of peptidolipin NA gave some N-acyl-L-Thr, showing that the acyl group was linked to the amino group of L-threonine. By reduction of peptidolipin a molecule of threonine was reduced to threoninol; under these conditions only the ester group was reduced. The carboxyl of a threonine residue was thus engaged in the lactone ring.

Saponification transformed neutral peptidolipin NA into peptidolipic acid of m.p. 107–110 °C, $[\alpha]_D - 10°$ (chloroform). The C-terminal amino acid of the latter was threonine, as shown above (*108*). Structure (**44**), $C_{50}H_{89}N_7O_{11}$, was proposed on the basis of these chemical reactions (*108*). This structure was confirmed by mass spectrometry (*109*).

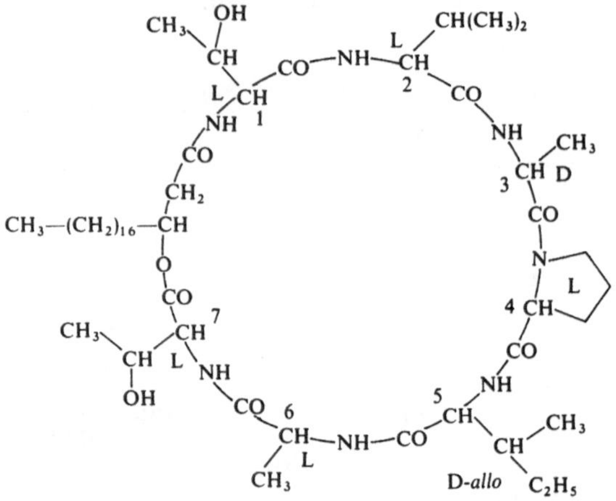

(**44**) Peptidolipin NA

The production of peptidolipin is not a general feature of strains of *Nocardia* as the search for peptidolipin in other strains of *N. asteroides* and *N. brasiliensis* was negative (*110*).

A ^{1}H NMR spectroscopic study of peptidolipin NA at 400 MHz has shown a rigid γ-turn in the sequence L-Pro-D-*allo*Ile which is stabilized by hydrogen bonds. The rest of the peptide ring is more flexible and plays a role in intermolecular interactions (*111*).

Countercurrent distribution gave peptidolipin NA as the major component of a peptidolipid fraction obtained from *N. asteroides* ATCC 9969. Two minor components were isolated. The first of these was "Val^6-peptidolipin NA" of m.p. 221–223 °C, $[\alpha]_D + 70°$ (chloroform). Chemical and mass spectrometric investigations showed that it differed from peptidolipin NA by replacement of the L-alanine residue in position 6 by a residue of L-valine. Beside 3D-hydroxyeicosanoic acid small amounts of the higher homologues C_{21} and C_{22} were observed (*112*). The second minor component, "Abu^1-peptidolipin NA", differed from peptidolipin NA by replacement of the threonine residue in position 1 by a residue of α-aminobutyric acid (Abu) which is a deoxythreonine. The lipid moiety was a mixture of the β-hydroxy fatty acids C_{20}, C_{21} and C_{22} (*113*).

3. N-Acylpeptides with a Lactone Ring not Involving the Hydroxyl Group of a Hydroxy Fatty Acid (N-Acylpeptolides)

In this group of compounds the peptide chain forms a lactone ring which involves the carboxyl group of the C-terminal amino acid and the hydroxyl group of a serine or threonine residue, thus giving what is called a peptolide. The acyl group is linked by an amide bond to the N-terminal amino acid.

3.1. N-Acyloctapeptides

3.1.1. Lipopeptins

Lipopeptin A is an antifungal antibiotic, $C_{54}H_{84}N_{10}O_{19}$, of m.p. 206–208 °C and $[\alpha]_D^{20}$ $-45°$ (methanol) which is produced by a strain of *Streptomyces* close to *Str. violaceochromogenes* (*114*). It was shown to inhibit the *in vitro* synthesis of peptidoglycan in *Escherichia coli* and of proteoheteroglycan in *Piricularia oryzae* cells (*115*). It was separated from a minor component designated as lipopeptin B which differed from lipopeptin A only in the structure of the fatty acid component (*114*).

Lipopeptin A is an acidic lipopeptide (p*Ka* 4.6). By acid hydrolysis it gave a fatty acid which was methylated with diazomethane and studied by GC/MS. It had the molecular weight of the methyl ester of a saturated C_{15} fatty acid. From the enhanced abundance of the ion *m/z* 199 $(M–C_4H_9)^+$ and the intensity of the peak at *m/z* 277 $(M–C_2H_5)^+$, higher than that of the peak at *m/z* 225 $(M–OCH_3)^+$, it was concluded that the acid was 12-methyl-tetradecanoic acid, a conclusion supported by comparison with an authentic sample.

The mixture of amino acids released by complete hydrolysis of lipopeptin A was fractionated by ion-exchange chromatography. Each amino acid was isolated and the optical rotation determined. Two moles of L-aspartic acid and 1 mole each of L-glutamic acid, L-serine, L-threonine, N-methyl-L-phenylalanine, L-*threo*-β-hydroxyglutamic acid and N-methyl-L-aspartic acid were obtained.

The presence of a lactone group was shown by isolation of an open-chain peptidolipid, $[\alpha]_D^{20}$ $-72°$ (methanol) after mild treatment with alkali. This derivative was hydrolyzed by carboxypeptidase with release of glutamic acid. The mixture of compounds obtained by refluxing a solution of lipopeptin A in dilute acetic acid was separated into free aspartic acid, N-acylthreonine, N-acyl-Thr-Asp and an hexapeptide [(**45**) in Scheme 8]. Structures of the N-acyl derivatives were established by EI mass spectrometry of the methyl esters.

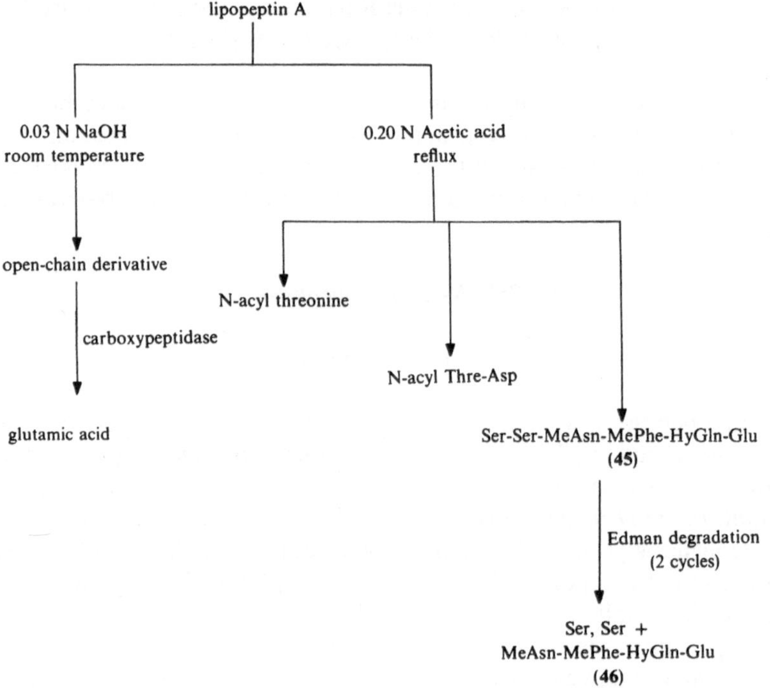

Scheme 8. Procedure used for establishing the structure of lipopeptin A

By Edman-dansyl degradations, the sequence Ser-Ser-MeAsn-MePhe for the N-terminal part of the hexapeptide was established. Tetrapeptide (**46**) obtained by elimination of the two serine residues from the hexapeptide was studied by NOE spectroscopy. Irradiation of the N-methyl protons of the N-methylphenylalanine residue and of the NH proton of glutamic acid indicated that N-methylasparagine and *threo*-β-hydroxyglutamine had their carboxamide groups in the β and γ positions respectively (*114*).

Chromic acid oxidation of lipopeptin A destroyed *threo*-β-hydroxy glutamic acid and serine, but not threonine. Therefore the lactone ring involved the hydroxyl group of threonine, a conclusion supported by comparing the ^{1}H NMR spectra of lipopeptin A and its open-ring derivative. Hence structure (**47**) was proposed for lipopeptin A. In lipopeptin B, the fatty acid was identified as 12-methyltridecanoic acid (*114*).

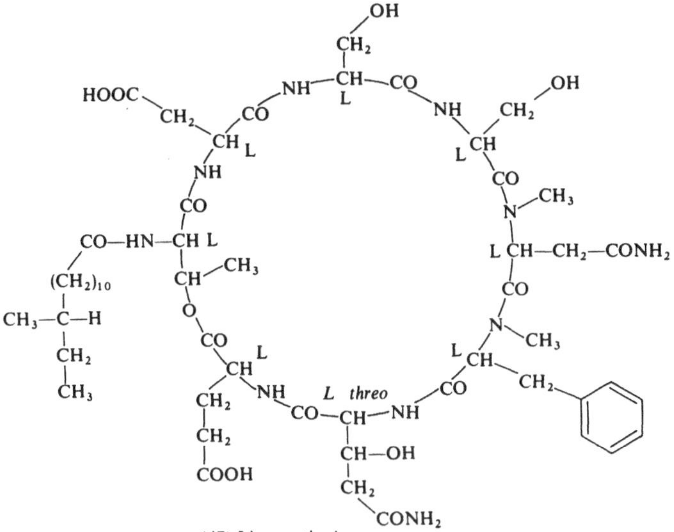

(**47**) Lipopeptin A

3.1.2. Neopeptins

One strain of *Streptomyces* produced a family of compounds with antifungal activity which inhibited the *in vitro* synthesis of proteoheteroglycan and 1,3-glycan synthetase in *Saccharomyces cerevisiae.*

The antibiotic fraction was separated into three closely related compounds, neopeptin A, $C_{53}H_{81}N_{11}O_{19}$, m.p. > 200° (dec.), $[\alpha]_D^{20}$ − 9° (methanol), neopeptin B, $C_{54}H_{83}N_{11}O_{19}$, m.p. > 200 °C (dec.), $[\alpha]_D^{20}$ − 28.5° (methanol) and neopeptin C, $C_{52}H_{79}N_{11}O_{19}$, m.p. 190–205°, $[\alpha]_D^{27}$ − 4° (methanol). All three compounds contained a basic group (pK*a* 9.5) and two acidic groups (pK*a* 5.9–4.0) (*116*, *117*).

By acid hydrolysis of neopeptin A seven amino acids were obtained, 2 moles of L-serine and 1 mole each of L-aspartic acid, L-*threo*-β-methylglutamic acid, L-*threo*-β-hydroxyglutamic acid, N-methyl-L-aspartic acid and N-methyl-L-phenylalanine. Several of these amino acids had previously been found in the lipopeptins (*114*).

Very mild treatment with alkali produced a ring-opened derivative, thus showing the presence of a lactone ring and allowing the release of β-methylglutamic acid by further treatment with carboxypeptidase.

By refluxing neopeptin with dilute acetic acid free aspartic acid, a tetrapeptide and N-acyl-Ser-Ser were obtained. The structure of the N-acyl-dipeptide was established by EI mass spectrometry of its methyl

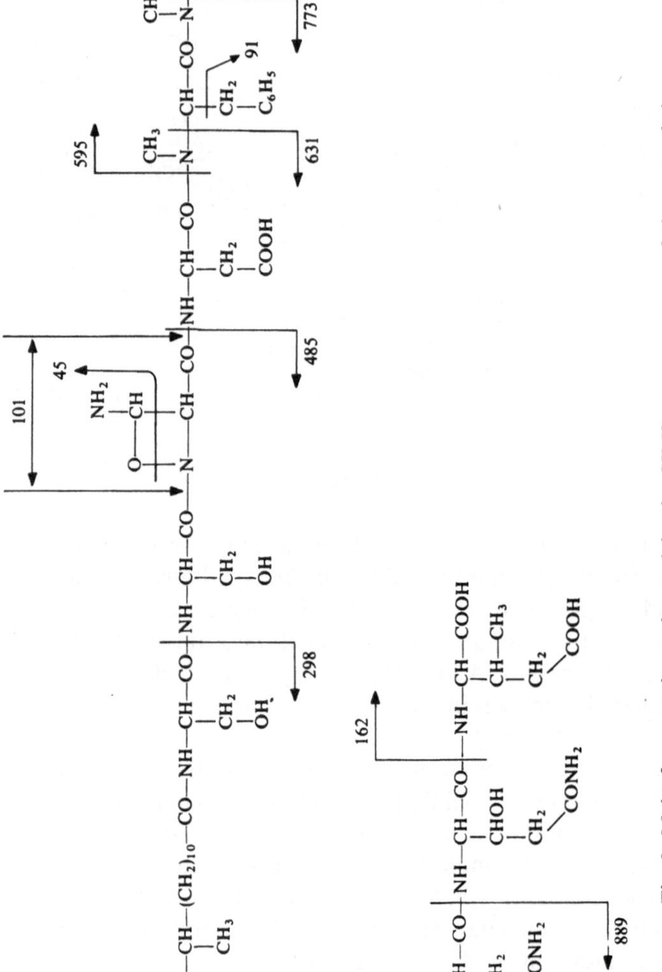

Fig. 3. Main fragment ions observed in the SIMS mass spectrum of the open-chain derivative of neopeptine A (from (*117*))

ester. The tetrapeptide contained N-methylaspartic acid, N-methyl-phenyl-alanine, *threo*-β-methylglutamic acid and *threo*-β-hydroxy-glutamic acid. Edman degradations performed on this tetrapeptide gave the sequence N-MePhe-N-MeAsn-HyGlu-MeGlu, the phenylthio-hydantoins having been identified by GC/MS.

The seven amino acids mentioned above linked together with 12-methyltridecanoic acid gave the formula $C_{50}H_{77}N_9O_{17}$, smaller than the formula determined by SIMS mass spectrometry for neopeptin A (m/z 1194 $(M + 1)^+$, m/z 1216 $(M + Na)^+$). The difference, $C_3H_4N_2O_2$, showed the presence of an eighth amino acid with a basic group, in agreement with the basic p*Ka* found for neopeptins.

Studies of the open-chain derivative of neopeptin A by SIMS mass spectrometry gave the results shown in Fig. 3. The structure of the labile basic amino acid was deduced as 3-amino-2-oxazetidine-4-carboxylic acid, in agreement with the fact that hydrogenation of neopeptin A followed by hydrolysis gave a small amount of 2,3-diaminopropionic acid. ^{13}C NMR spectra were consistent with the presence of a four-membered ring. Therefore structure (**48**) was proposed for neopeptin A.

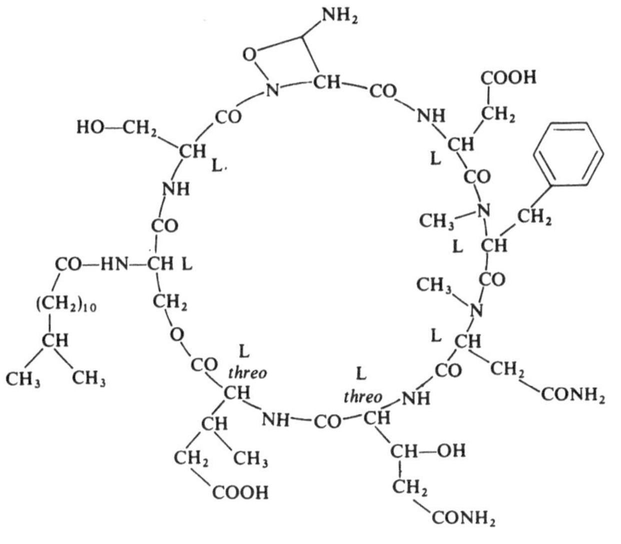

(**48**) Neopeptin

Neopeptin B contained 12-methyltetradecanoic acid as the fatty acid moiety and neopeptin C, 11-methyldodecanoic acid (*117*).

From a careful study of neopeptin A by ^1H-NMR spectroscopy, 2D-NMR using COSY and NOESY methods at 400 MHz, two conformations were deduced with two hydrogen bonds involving the NH and CO groups of the residues of aspartic and β-hydroxyglutamine (*117*).

Close similarities can be noted between the structures of lipopeptins and neopeptins (formulas **47** and **48**).

3.2. N-Acylnonapeptide: Viscosin

A peptidolipid exhibiting antibiotic properties against mycobacteria and viruses produced by *Pseudomonas viscosa* was obtained as a crystalline compound of m.p. 270–273 °C, $[\alpha]_D^{29}$ − 168° (ethanol). The N-3D-hydroxydecanoyl-hexapeptide structure (**49**) proposed originally (*118*) was later shown to be incorrect because the synthetic product was different from natural viscosin (*119*).

CH_3—$(CH_2)_6$—CHOH—CH_2—CO—L. Leu-Gly-L. Ser-D. Val-L. Thr-L. Leu-OH

(**49**)

Mild saponification of viscosin gave viscosic acid, m.p. 170–175 °C, $[\alpha]_D$ − 10° (ethanol) whose infrared spectrum no longer displayed an ester or lactone carbonyl band (1739 cm^{-1}) present in viscosin, but had retained the same amino acid composition Leu, Ile, Val, Glu, Ser, *allo*Thr in the molar ratio 3:1:1:1:2:1. Mass spectrometry of the permethylated derivative of viscosic acid gave the amino acid sequence. From the mass spectrum of viscosic acid dimethyl ester and p*Ka* measurements it was concluded that glutamic acid is bound through its α-carboxyl group. Viscosin and its methyl ester, prepared by the action of diazomethane, were perdeuteriomethylated; comparison of the mass spectra of the two derivatives showed that the γ-carboxyl group of glutamic acid was free in viscosin and that only the carboxyl group of the C-terminal amino acid, isoleucine, could be engaged in the lactone ring.

Viscosin and viscosic acid were oxidized with chromic acid and the amino acids obtained after hydrolysis of the oxidation products were analyzed. *Allo*threonine was found only in the products arising from viscosin, showing that the hydroxyl group of the amino acid was protected by esterification with the carboxyl group of isoleucine in the antibiotic.

The configuration of the amino acids was determined by using L-glutamic acid decarboxylase and D-amino acid oxidase. The complete structure of viscosin was thus established as (**50**) (*120*).

Solid phase synthesis of viscosin confirmed structure (**50**). In order to obtain the lactone ring in the right position the synthesis was performed through the intermediate (**51**) (*121*).

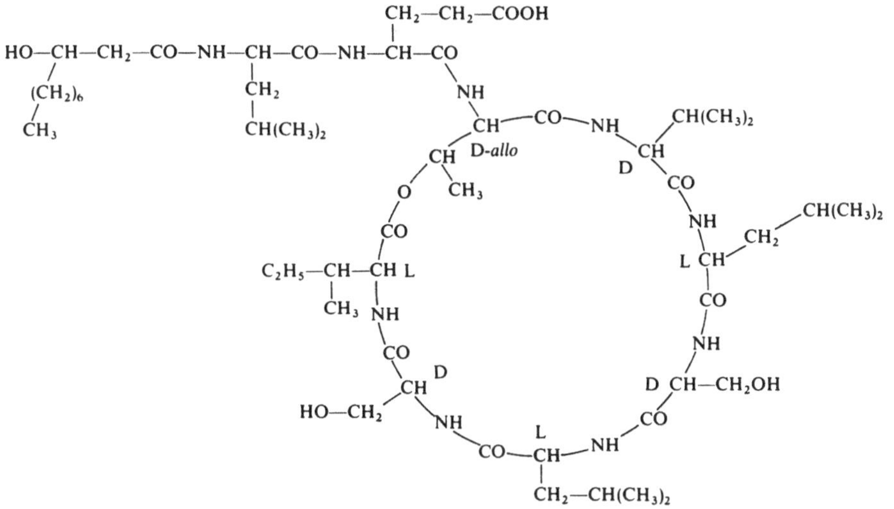

```
                                          CH2—CH2—COOH
                                          |
HO—CH—CH2—CO—NH—CH—CO—NH—CH—CO
   /              |                        \
 (CH2)6          CH2                        NH
  /               |                          \
 CH3             CH(CH3)2                    CH—CO—NH       CH(CH3)2
                                            /              \  /
                                          CH  D-allo        CH
                                         / \              D  \
                                        O   CH3               CO
                                       /                        \
                                     CO                          NH        CH(CH3)2
                                     /                             \   CH2
                     C2H5—CH—CH L                               L CH
                          |   |                                   |
                         CH3  NH                                  CO
                               \                                  /
                               CO                                NH
                                 \   D                          /
                                  CH                      D CH—CH2OH
                                 /  \                       /
                          HO—CH2     NH                   CO
                                       \      L          /
                                        CO—CH—NH
                                             |
                                            CH2—CH(CH3)2
```

(50) Viscosin

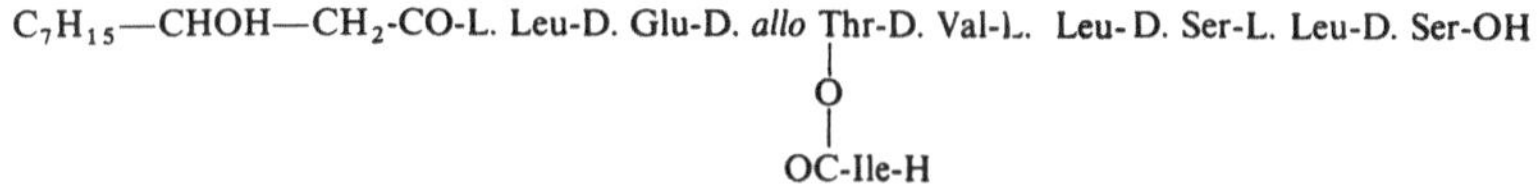

C_7H_{15}—CHOH—CH_2-CO-L. Leu-D. Glu-D. *allo* Thr-D. Val-L. Leu-D. Ser-L. Leu-D. Ser-OH

```
                 |
                 O
                 |
              OC-Ile-H
```

(51)

3.3. N-Acyldecapeptides: Imacidines

A strain of *Streptomyces* recognized as *Str. olivaceus* and isolated from a sample of soil obtained in Turkey was found to produce a family of peptidolipidic antibiotics along with prodigiosin and metacycloprodigiosin. These antibiotics interfered with the biosynthesis of murein and only the growth of actinomycetes was inhibited (*123*).

The crude antibiotic called imacidin was fractionated by chromatography on a cation-exchange resin into several components. All of them had the same amino acid composition, but differed from each other by the length of the fatty acid component (C_{11}–C_{15}). Imacidin C, the main component of the mixture, contained 3-hydroxyisotridecanoic acid (*123*).

After partial hydrolyses and the usual degradative methods, structure (**52**) was proposed (*124*).

As peptidolipids having a fatty acid component with less than 10 carbon atoms are not within the scope of this review, the N-acylundecapeptide *brevistin* (*122*) will not be discussed here.

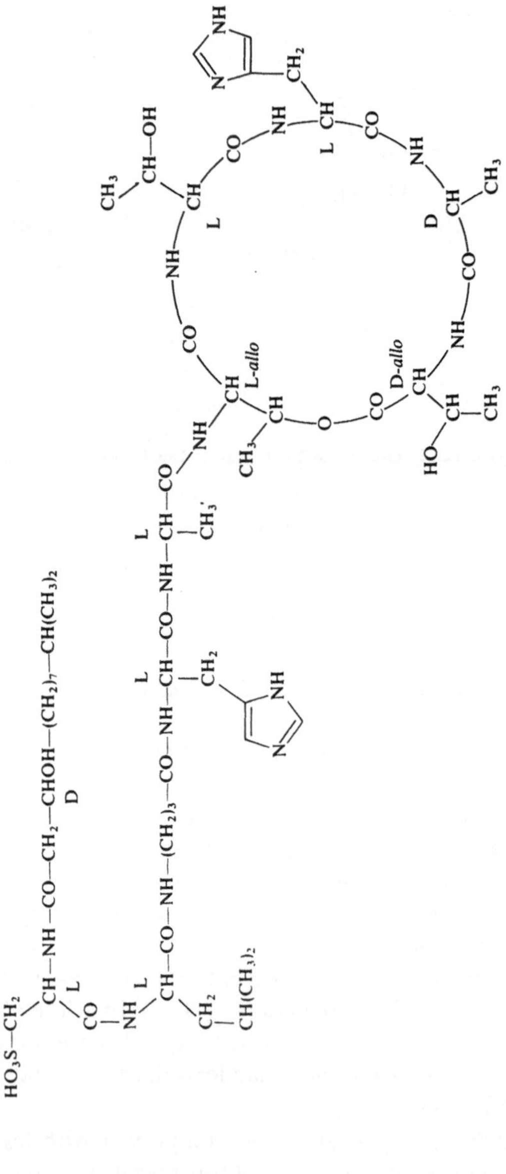

(52) Imacidin C

3.4. *N-Acyltridecapeptides: Antibiotic A21978 C*

A strain of *Streptomyces roseosporeus* produced a group of antibiotics designated A21978, fractions A, B, C, D and E, in the culture medium. The major component, A21978 C, was highly active against Gram positive bacteria and was shown by HPLC to consist of at least six closely related antibiotic substances. The major components were C_1 $C_{73}H_{103}N_{17}O_{26}$, C_2 $C_{74}H_{105}N_{17}O_{26}$ and C_3 $C_{75}H_{107}N_{17}O_{26}$ (*125*).

Fatty acids, first detected because of the presence of strong methyl and methylene signals at δ0.98 and δ1.1 in the ^{1}H NMR spectra of C_1, C_2 and C_3, were isolated by acid hydrolysis and identified by GC/MS of their methyl esters. They were respectively C_{11}, C_{12} and C_{13} fatty acids. One of the fatty acids was attached to the N-terminal part of the peptide moiety.

Potentiometric titration showed pK*a* values of ~ 5.8, 7.6 and ~12 which were consistent with the presence of several carboxylic acid groups and of an aromatic and an aliphatic amino group. Because of the presence of these free ionisable groups, A21978 C possesses polar character and is soluble in water and lower alcohols.

The mixture of amino acids obtained by acid hydrolysis was fractionated by ion-exchange chromatography. An unusual new amino acid was isolated. It was converted into its phenylthiohydantoin (PTH) derivative, the EI mass spectrum of which [M^+ 278, *m/z* 260 (M-18), 232 (M-46), 218 (M-60)] was identical with that of the PTH derivative of 3-methylglutamic acid. Thanks to the use of glutamine synthetase and L-glutamic acid decarboxylase, the *threo* configuration and the L-series of this amino acid could be established. The configuration of the other amino acids, L-Asp, L-Trp, L-kynurenine, was studied either by the use of D- or L-amino acid oxidase or by optical rotation measurements. The presence of kynurenine explained the fluorescence of A21978 C under long wave length U.V. light on thin layer chromatograms.

The presence of a lactone group in A21978 C was detected as follows. A band at 1740 cm^{-1} in the infrared spectrum was shifted to 1720 cm^{-1} when the antibiotic was treated with dilute base at room temperature. Chromic acid oxidation of A21978 C destroyed serine but not threonine. These data supported the presence of a lactone ring involving the hydroxyl group of threonine (*125*).

Treatment of the antibiotic with carboxypeptidase Y showed the absence of a C-terminal amino acid, while L-kynurenine was released when the antibiotic was pretreated with base to open the lactone ring. Therefore the lactone ring involved the carboxyl group of kynurenine as well as the hydroxyl group of threonine.

From a partial hydrolysate of A21978 C an ether-soluble fraction was isolated, which was methylated with diazomethane. The mass spectrum of this material (prepared from fraction C_1) gave a molecular ion at *m/z* 386 and ion fragments at *m/z* 201 and 130, and was identical with the mass spectrum of authentic N-undecanoyltryptophan methyl ester. Studies of the N-acyltryptophan prepared from fractions C_1, C_2 and C_3 by EI mass spectrometry of the methyl esters and by CID mass spectrometry of the carboxylate ions in the negative mode showed that C_1 contained 9-methyldecanoic acid (anteiso-C_{11}), C_2 10-methylundecanoic acid (iso-C_{12}) and C_3 10-methyldodecanoic acid (anteiso-C_{13}) (*125*).

The mixture of A21978 C compounds was deacylated by fermentation with *Actinoplanes utahensis* and a single peptide moiety was obtained (*126*). Preliminary experiments involving Edman degradation of the base-pretreated peptide with an open lactone ring gave a low yield of cleavage products explained by the presence of Asn residues and $\alpha \rightarrow \beta$ amide bond rearrangement due to the base treatment. By studying the intact peptide, sequence (**53**) was established.

Shorter peptides obtained by partial acid hydrolysis were separated by ion-exchange chromatography and studied by EI mass spectrometry of their N-trifluoroacetyl methyl esters and the reduction products obtained after hot treatment with $LiAlD_4$. Consequently, structure (**54**) was proposed for the A21978 C antibiotics, each member of the group differing from the others only by the structure of the fatty acyl moieties (*125*).

Analogous lipopeptides were prepared from A21978 C by enzymatic hydrolysis using *Actinoplanes utahensis* and reacylation after suitable protection of the peptide, for evaluation of their antibiotic properties (*127*). The *n*-decanoyl analogue had the lowest level of toxicity and was called daptomycin.

Monolayer characteristics of A21978 C and of daptomycin were studied by varying the nature of the cations contained in the subphase. The mean area of the uncompressed molecules was about 220–240 Å^2 which is compatible with the size of the peptide cycle lying at the interface (*128*).

3.5. N-Acyltetradecapeptide: Stendomycin

The antifungal antibiotic stendomycin was isolated from *Streptomyces antimycoticus* as a family of closely related components which differed from each other by their fatty acid constituents (*129*). Two fractions, A ($[\alpha]_D$ − 92.4° (ethanol)) and B ($[\alpha]_D$ − 83.4° (ethanol)) were described, but the difference between them was found to be in the anion

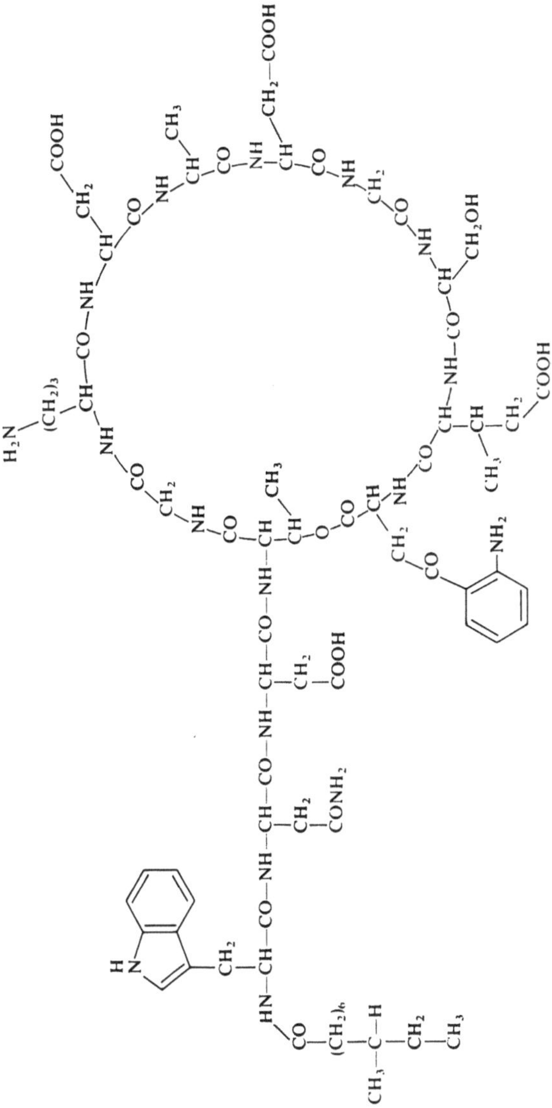

H-Trp-Asn-Asp-Thr-Gly
|
O—R

(53)

(54) Antibiotic A21978 C_1

associated with the single cationic center of the molecule (*130*). Because of its properties as a salt stendomycin does not behave as a true lipid and has a relatively polar character.

The mixture of fatty acids obtained by acid hydrolysis of stendomycin consisted of 12-methyltridecanoic acid and 11-methyldodecanoic acid as major components together with some lower homologues (*131*). Fourteen amino acids were characterized (*132*). Their estimation gave the following composition: Ala_1, Gly_1, Ser_1, Pro_1, Val_3, *allo*Ile_2, *allo*Thr_2, N-Me Thr_1, 1 mole of dehydro-α-aminobutyric acid and 1 mole of a basic amino acid which is an N,N′-dimethyl derivative of a "cyclic arginine" (**55**). Dehydro-α-aminobutyric acid gave rise to α-ketobutyric acid by acid hydrolysis of stendomycin and α-aminobutyric acid by hydrolysis of dihydrostendomycin (*133*). NMR spectroscopy helped to determine the structures of these two unusual amino acids. Optical rotation measurements of the isolated amino acids showed that both *allo*threonines and both *allo*isoleucines, alanine and one valine (out of three) belonged to the D-series of amino acids.

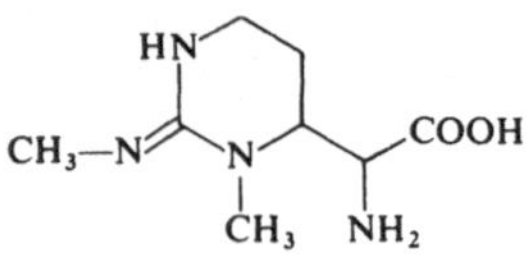

HN

CH_3—N N COOH

CH_3 NH_2

(**55**)

The infrared spectrum of stendomycin exhibited a band at 1738 cm^{-1} attributed to a lactone group. In keeping with this, mild alkaline hydrolysis converted stendomycin to stendomycic acid. On chromic acid oxidation of stendomycin, only one of the *allo*threonine residues remained intact while the other *allo*threonine, the N-methylthreonine and the serine residues disappeared from the hydrolysates. This result showed that the hydroxyl group of one *allo*threonine residue participated in the formation of the lactone. Reduction of stendomycin with $NaBH_4$ reduced only the "cyclic arginine" derivative, indicating that it was the C-terminal amino acid which was involved in formation of the lactone ring.

Partial hydrolysis of stendomycin by concentrated hydrochloric acid at room temperature yielded a mixture of peptides which were separated by ion-exchange chromatography. The isolation of N-acyl proline and N-acyl Pro-N-MeThr identified the N-terminal end of the peptide chain. Partial hydrolysis or methanolysis of dihydrostendomycin preferentially cleaved the peptide chain after the α-aminobutyric acid residue, allowing the isolation of two peptides containing the six or seven first amino acids from the N-terminal end.

On the basis of these results, structure (**56**) was proposed (*130*). It is in complete agreement with the results of a mass spectrometric study, which also indicated that about 30% of *allo*isoleucine-13 was replaced by valine (*134*).

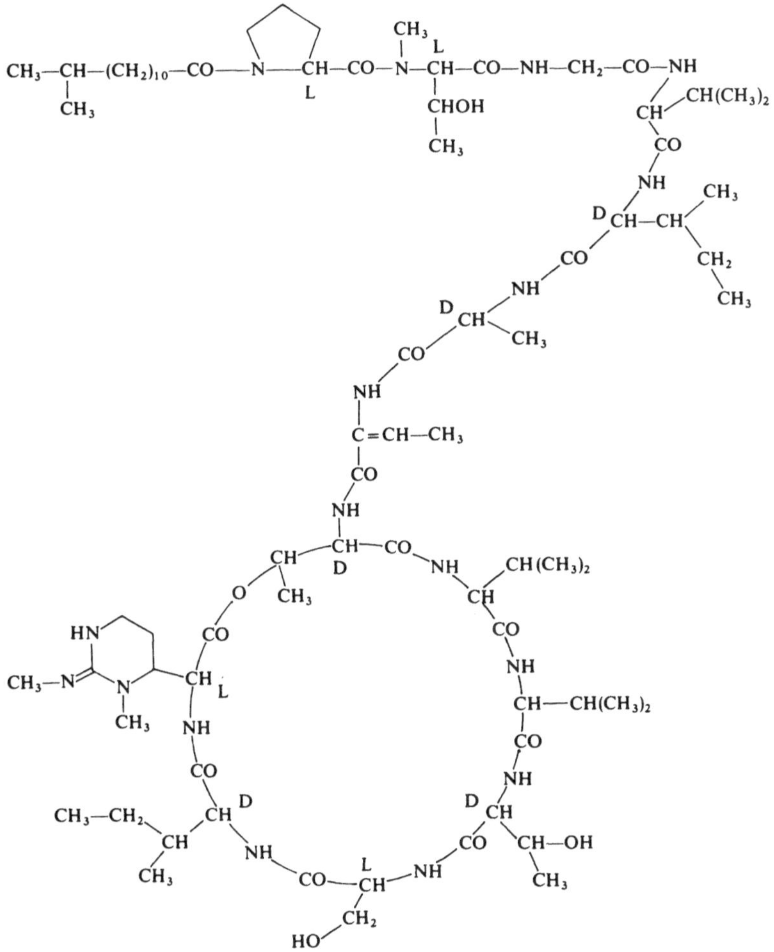

(**56**) Stendomycin

The conformation of stendomycin was studied by circular dichroism. In trifluoroethanol the circular dichroism pattern was almost identical with that of left-handed α-helical poly-D-glutamic acid in water at pH 3.9; the amplitude was half that of poly-D-glutamic acid. The presence of α-helical zones in stendomycin was supported by the occurrence of an amide I band at 1659 cm^{-1} in the infrared spectrum, while gramidine S, known for its β-structure, has a band at 1637 cm^{-1} (*135*).

4. Peptidolipids with a Lactam Ring

A group of peptidolipids produced by strains of *Bacillus subtilis* is characterized by the presence of β-amino fatty acids as lipid components. The peptide chain forms a macrocyclic lactam by reaction between the amino group of the fatty acid and the carboxyl group of the C-terminal amino acid. These compounds are isolated from the culture medium of the bacteria and have mainly antifungal properties.

4.1. Iturins

B. subtilis produced a peptidolipid fraction named *iturin* because the bacterial strain had been isolated from the soil in a village called Ituri in Zaire. By chromatography on silicic acid three components (A, B and C) were obtained (*136*). In fact the mixture was more complex as it proved possible to resolve iturin A into seven components by HPLC (*137*).

The two major components of iturin A differed only in the nature of the lipid moiety. By hydrolysis of iturin A, about 20% of a lipid substance giving a positive reaction with ninhydrin was isolated. Its behaviour in acid or alkaline medium showed that it was an amino acid. Its mass spectrum exhibited molecular peaks at *m/z* 257 ($C_{14}H_{28}(NH_2)$–COOH) and *m/z* 243 ($C_{13}H_{26}(NH_2)$–COOH). The GC/MS spectra of the N-acetyl methyl ester of the two homologous amino fatty acids exhibited common peaks at *m/z* 144 corresponding to the ion CH_3–CO–$\overset{+}{N}H$=CH–CH_2–$COOCH_3$ and *m/z* 102 ($\overset{+}{N}H_2$=CH–CH_2–$COOCH_3$). The other part of the molecule gave two peaks at *m/z* 240 ($C_{12}H_{25}$–CH=$\overset{+}{N}H$–CO–CH_3) and *m/z* 226 ($C_{11}H_{23}$–CH=$\overset{+}{N}H$–CO–CH_3). Thus two β-amino acids with 14 and 15 carbon atoms were present and were called iturinic acids.

Five α-amino acids were obtained by complete acid hydrolysis, aspartic acid and glutamic acids, tyrosine, proline and serine in the molar ratio 3:1:1:1:1. Their configuration was determined enzymatically; two out of three residues of aspartic acid and the residue of tyrosine belonged to the D-series.

Iturin A behaved as a neutral compound on paper electrophoresis and alkalimetric titration; it was thus likely that the aspartic and glutamic acid residues were present in the form of asparagine and glutamine. In order to demonstrate this, iturin A was dehydrated and reduced following the procedure of RESSLER and KASHELIKAR (*86*). In this manner asparagine and glutamine were transformed into 2,4-diaminobutyric acid and ornithine respectively.

Chromic acid oxidation resulted in disappearance of serine and tyrosine; likewise no ester absorption band was detected in the infrared spectrum of iturin A, thus precluding the presence of a lactone group.

Partial acid hydrolysis gave an amide of the β-amino fatty acid and serine, with an N-terminal position for the serine residue. The structures of several other peptides were established by chemical means and by mass spectrometry of their permethylated derivatives. From these results, the complete structure of iturin A was established as (**57**) (*138*). Circular dichroism of derivatives of the β-amino fatty acids (iturinic acids) showed their configurations to be *R* (*139*).

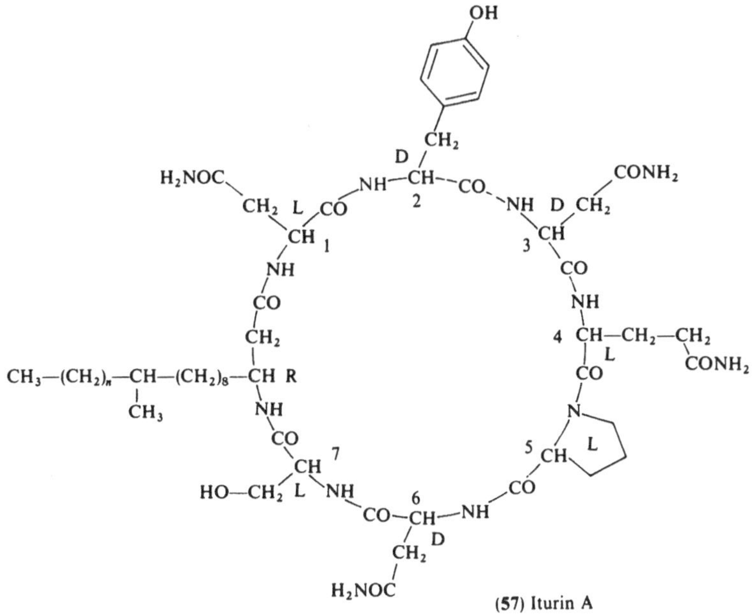

(**57**) Iturin A
$n = 0$ or 1

Structure (**57**) was in agreement with the results of mass spectrometric investigations of iturin A as a permethylated derivative. The interpretation of the mass spectra was made easier by the presence of pairs of peaks differing by 14 mass units due to the presence of the two homologous β-amino fatty acids (*138*).

Studies by ^{1}H- and ^{13}C-NMR spectroscopy of the conformation of iturin A in solution in deuterated DMSO showed the presence of hydrogen bonds between the NH group of the L-serine residue and the CO group of the L-glutamine residue, and between the NH group of L-glutamine and the CO group of an L-asparagine residue. The results favoured a relatively rigid structure with two β-turns and a small inner cavity (< 2 Å diameter), a type of structure similar to that generally found in ionophores. The α-amino acid side-chains had a larger degree of freedom allowing possible interactions with components of the bacterial envelope (*140*, *141*).

From the culture medium of a strain of *B. subtilis* another sample of iturin was isolated which was called iturin A_L. It was resolved into six components by HPLC. FAB mass spectrometry gave three major molecular peaks: MH^+ 1043, 1057 and 1071 showing that iturin A_L is mainly a mixture of homologues $C_{48}H_{74}N_{12}O_{14}$ (M 1042) and $C_{49}H_{76}N_{12}O_{14}$ (M 1056). The lipids obtained by hydrolysis and analyzed by GC/MS were identified as β-amino fatty acids; the mass spectra of their trifluoroacetyl methyl esters exhibited a base peak at *m*/*z* 198. Their amino acid composition was the same as that of iturin A (see Table 7). The name A_L was said to have been used because of the high content of long chain β-amino fatty acids (*142*).

As mentioned earlier, the iturin fraction was separated into a major component, iturin A, and small amounts of related compounds. Among them, only iturin A exhibited antifungal activity.

Iturin C (about 1 to 3% of the crude iturin fraction) contained one free carboxyl group as shown by its electrophoretic behaviour and by alkalimetric titration. The carboxyl group was reduced by diborane, causing the disappearance of one L-aspartic molecule in the hydrolyzate and the simultaneous appearance of one mole of homoserine. Mass spectrometry of the permethylated derivative of iturin C allowed localization of the aspartic acid residue in position 1; thus iturin C is Asp^1-iturin A (*143*).

Iturin D also contained one free carboxyl group in place of a carboxamide group and in iturin E, one carboxamide group was replaced by carboxymethyl group (*144*).

The peptide moiety of iturin A was synthesized with the purpose of preparing analogues differing in the length of the hydrophobic part (*145*).

4.2. Bacillomycins

Other strains of *B. subtilis* produced antibiotics which were studied under the name of bacillomycins in spite of the fact that the designation *mycin* should be reserved for antibiotics produced by *Streptomyces*.

The structures of three bacillomycins D, F and L have been completely established as (**59**), (**58**) and (**60**) respectively. All of them contained the characteristic iturinic acids and seven α-amino acids. The sequence of amino acids was studied using chemical methods, in particular Edman degradation of the product obtained by specific cleavage of the peptide moiety with N-bromosuccinimide. Confirmatory results were obtained by chemical ionisation (isobutane) mass spectrometry and various techniques of NMR spectroscopy. The exact molecular weights of the various species were obtained by FAB mass spectrometry (*146*, *147*) (see Table 7).

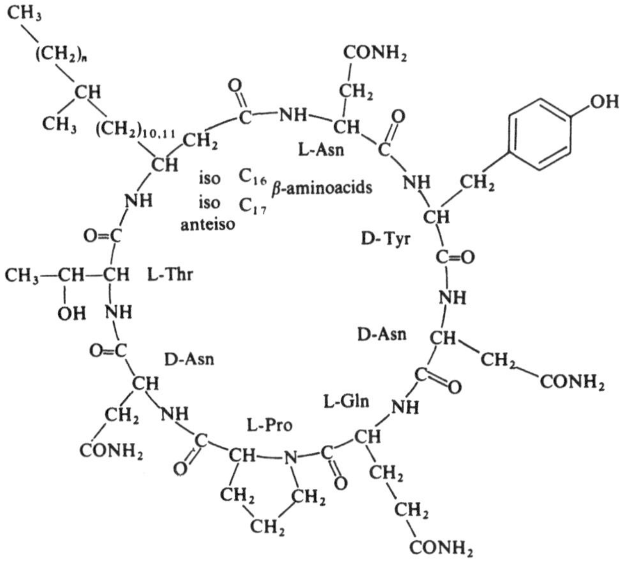

(**58**) Bacillomycin F $n = 0$ or 1

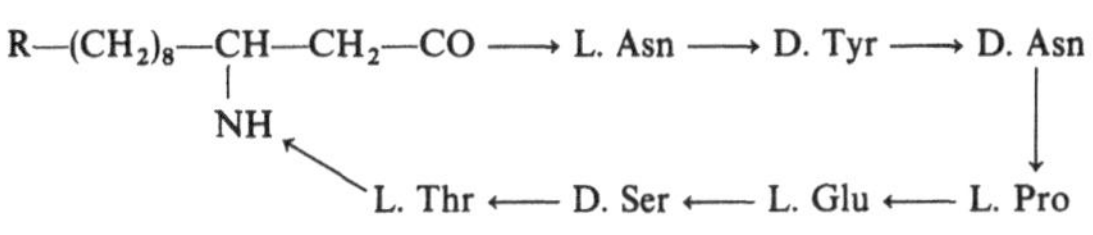

R—$(CH_2)_8$—CH—CH_2—CO ⟶ L. Asn ⟶ D. Tyr ⟶ D. Asn ↓ L. Pro ⟶ L. Glu ⟶ D. Ser ⟶ L. Thr ⟶ NH

(**59**) Bacillomycin D

R—$(CH_2)_8$—CH—CH_2—CO ⟶ L. Asp ⟶ D. Tyr ⟶ D. Asn ↓ L. Ser ⟶ L. Gln ⟶ D. Ser ⟶ L. Thr ⟶ NH

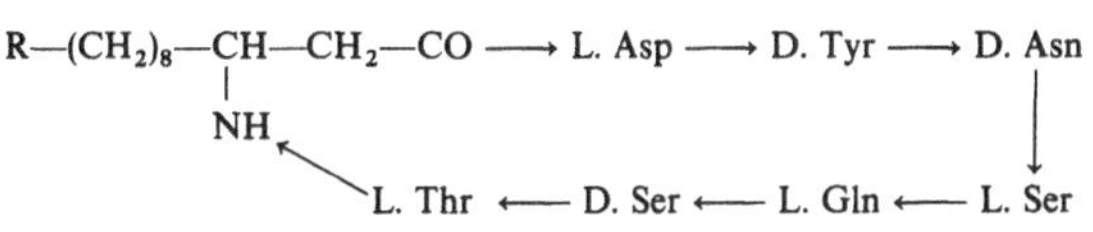

(**60**) Bacillomycin L

$$
\begin{array}{l}
R\text{—}(CH_2)_{10}\text{—}CH\text{—}CH_2\text{—}CO \longrightarrow L.\,Asn \longrightarrow D.\,Tyr \longrightarrow D.\,Asn \\
\qquad\qquad\quad | \qquad\qquad\qquad\qquad\qquad\qquad\qquad\qquad\qquad\qquad \downarrow \\
\qquad\qquad\quad NH \nwarrow L.\,Asn \longleftarrow D.\,Ser \longleftarrow L.\,Pro \longleftarrow L.\,Gln
\end{array}
$$

(**61**) Mycosubtilin

$R = CH_3\text{—}CH_2\text{—}CH_2\text{—}$, $CH_3\text{—}CH(CH_3)\text{—}$, $CH_3\text{—}CH(CH_3)\text{—}CH_2\text{—}$, $CH_3\text{—}CH_2\text{—}CH(CH_3)\text{—}$

More recent studies have shown the presence, besides bacillomycin F $C_{52}H_{82}N_{12}O_{14}$ (later renamed bacillomycin F_a), of two closely related compounds, bacillomycin F_b $C_{52}H_{81}N_{11}O_{15}$ in which one carboxamide group is replaced by a free carboxyl group, and bacillomycin F_c $C_{52}H_{80}N_{10}O_{16}$ in which two carboxamide groups are replaced by two free carboxyl groups (*148*).

4.3. Mycosubtilin

Another antifungal agent is produced by another strain of *B. subtilis* as a mixture of peptidolipids. Two homologous peptidolipids are the major components of the mixture; they contain β-amino fatty acids easily characterized by a strong peak at *m/z* 240 in the mass spectrum of the N-trifluoroacetyl butyl ester (*149*).

Table 7. *Composition of the Peptidolipids of the Iturin Group*

$$
\begin{array}{l}
\qquad\qquad\qquad\quad 1 \qquad\qquad\quad 2 \qquad\qquad 3 \\
R\text{—}CH\text{—}CH_2\text{—}CO\text{—}L.Asx \longrightarrow D.Tyr \longrightarrow D.Asn \\
\quad\; | \\
\quad NH \\
\quad\; | \\
\quad X_7 \longleftarrow X_6 \longleftarrow X_5 \longleftarrow X_4 \swarrow
\end{array}
$$

	x L	X_4 L	X_5 L	X_6 D	X_7 L	Major β-amino fatty acids
Iturine A	Asn	Gln	Pro	Asn	Ser	i-C_{14}, a-C_{15}
Iturine A_L	Asn	Gln	Pro	Asn	Ser	n-C_{16}, i-C_{16}
Iturine C	Asp	Gln	Pro	Asn	Ser	i-C_{14}, a-C_{15}
Bacillomycine D	Asn	Pro	Glu	Ser	Thr	n-C_{14}, i-C_{15}
Bacillomycine L	Asp	Ser	Gln	Ser	Thr	n-C_{14}, i-C_{15}
Bacillomycine F	Asn	Gln	Pro	Asn	Thr	i-C_{16}, i-C_{17}
Mycosubtiline	Asn	Gln	Pro	Ser	Asn	n-C_{16}, i-C_{16}, i-C_{17} a-C_{17}

By using FAB mass spectrometry, two dimensional ^{1}H-NMR spectroscopy and chemical degradations, structure (**61**) was established. This structure differs from that of the iturins by inversion of the two amino acid residues at the C-terminal end of the peptide moiety. D-Ser-L-Asn in mycosubtilin replaces D-Asn-L-Ser in the iturins (*150*). The β-amino fatty acids are *anteiso*-C_{17} and *iso*-C_{16}-iturinic acids (see Table 7).

4.4. General Comments on the Peptidolipids of the Iturin Group

This group is characterized by the presence of β-amino fatty acids, the iturinic acids. These acids are derivatives of the following fatty acids: *n*-C_{13}, *n*-C_{14}, *anteiso*-C_{15}, *iso*-C_{15}, *n*-C_{15}, *iso*-C_{16}, *n*-C_{16}, *anteiso*-C_{17} and *iso*-C_{17}. The ratios of the different amino fatty acids depend on the nature of the culture medium which can afford different precursors.

In addition to the differences in amino fatty acid composition, the various lactams that have been isolated exhibit some differences in α-amino acid composition (Table 7). At first it can be observed that the N-terminal part of the peptide moiety (sequence $1 \rightarrow 3$) is always constant except for the replacement of Asn and Asp. Frequent modifications are observed in the sequence of amino acids $4 \rightarrow 7$ which is the variable part of the molecule. What is most interesting is the fact that when a replacement occurs or an inversion in the sequence takes place the positions of the D or L residues remain quite stable. The most illustrative example is provided by iturin A and mycosubtilin with the respective sequences D-Asn-L-Ser and D-Ser-L-Asn (see Table 7). The configurations of the amino acids are more important than their structures for the role that they have to play in the bacterial environment.

At the end of this review, another compound containing a β-amino fatty acid, puwainaphycin C, will be mentioned.

4.5. Octapeptins

A strain of *Bacillus circulans* was found to produce a family of basic water-soluble lipopeptides designated as octapeptins which exhibited antibiotic activity against Gram negative bacteria. These lipopeptides show similarities to the polymyxins which are not discussed here because of the insufficient length of their acyl chain. All these lipopeptides have a high content of 2,4-diaminobutyric acid (DAB), five DAB residues out of eight amino acids in the case of the octapeptins.

These lipopeptides are devoid of β-amino fatty acids and are not true lactams. They consist of a cyclic heptapeptide with an *exo*-amino acid

Table 8. *Structures of the Members of the Octapeptin Family* (for the type of fatty acid, see text)

```
              γNH2               γNH2
fatty acyl ——→ D.DAB ——→ L.DAB——→ L.DAB ——→ X ——→ Y
                           ↑                      ↙
                         L.Leu ←—— L.DAB ←—— L.DAB
                                   γNH2      γNH2
```

Compound	Type of fatty acid	X	Y	$[\alpha]_D^{23}$ (hydrochloride)
A_1	1	D. Leu	L. Leu	
A_2	2	"	"	
A_3	3	"	"	
B_1	1	D. leu	L. Phe	– 20.5° (DMSO)
B_2	2	"	"	
B_3	3	"	"	
C_1	4	D. Phe	L. Leu	– 65.6° (0.5 N HCl)

residue, the amino group of which bears a fatty acyl group (see Table 8). The fatty acid components were isolated and identified. Four kinds of fatty acids were found, in the lipopeptides 1 8-methyldecanoic acid, in the lipopeptides 2 8-methylnonanoic acid, in the lipopeptides 3 *n*-decanoic acid and in lipopeptide C_1 6-methyloctanoic acid as in the polymyxins.

Table 8 shows structures of seven members of the octapeptin family (*151*).

4.6. *Antibiotic EM 49*

A strain of *B. circulans* produces a broad spectrum antibiotic designated as EM 49. This antibiotic is a mixture of closely related lipopeptides which contain five DAB residues out of eight amino acids like the octapeptins. The structure of the peptide moiety of EM 49 is identical with that of octapeptin C_1 (Table 8). The fatty acids of EM 49 are 3-hydroxy fatty acids with 10 or 11 carbon atoms. They were identified as 3-hydroxy-8-methylnonanoic acid, 3-hydroxydecanoic acid and 3-hydroxy-8-methyldecanoic acid (*152*).

5. Mass Spectrometry of N-Acylpeptides

The application of mass spectrometry to structure determination of the peptidolipids has played an important role in development of this technique for peptide analysis. The interpretation of the mass spectra of peptidolipids was made easier because of the natural labelling of the N-terminal part of the peptide chain by an acyl moiety which is a mixture of closely related homologues. In spite of the advances made simultaneously by chemical methods, mass spectrometry retains several important advantages particularly because of the small amounts of material required for analysis and the facile cleavage of sterically hindered peptide bonds such as Ile–Ile.

Peaks corresponding to cleavage of the peptide bonds in the EI mass spectra of peptide derivatives are found as shown below:

$$\left[\cdots\mathrm{NH{-}\underset{\displaystyle R_x}{CH}{-}CO{-}NH{-}\underset{\displaystyle R_y}{CH}{-}COOCH_3}\right]^{+} \longrightarrow \left[\cdots\mathrm{NH{-}\underset{\displaystyle R_x}{CH}{-}C{\equiv}O}\right]^{+} + \mathrm{NH{-}\underset{\displaystyle R_y}{CH}{-}COOCH_3}$$

In higher oligopeptides, peaks due to the acylium ions are often the most intense. They appear as groups of two or three peaks differing by 14 or 28 mass units because of the labelling of the N-terminal part by the mixture of fatty acid homologues. The presence of the fatty acid moiety has one more advantage, in that it shifts the useful peaks to a less crowded, higher mass region so that clearer mass spectra are obtained (*153*). The appearance of the spectrum is the result of sequential splitting of the peptide chain, one amino acid residue after the other, from the C-terminal end of the molecular ion. As an example, the EI mass spectrum of the peptidolipid isolated from *M. paratuberculosis* is shown in Fig. 4.

The analysis of N-acylpeptides with a large number of amino acid residues by EI mass spectrometry requires that intermolecular hydrogen bonds be broken or avoided altogether. This is done by methylating the

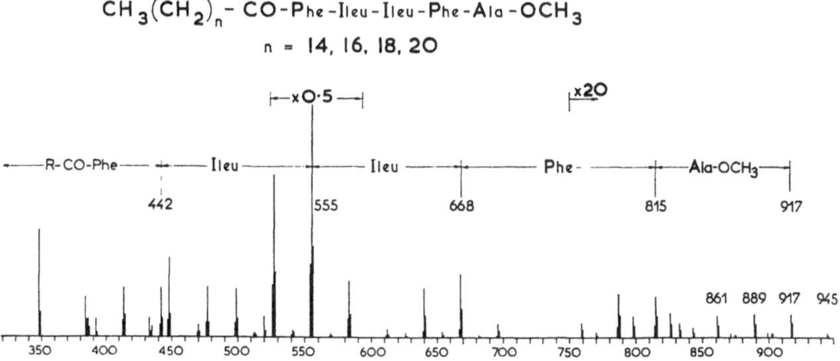

Fig. 4. EI-mass spectrum of the peptidolipid isolated from *Mycobacterium paratuberculosis* (*76*)

NH groups of the peptide bonds employing, for example, methyl iodide and silver oxide in dimethylformamide (*154*) or HAKOMORI's method using methylsulfinyl carbanion (*155*). Methylation of peptides not only increases their volatility, but also improves cleavage of the peptide bonds, thus giving a clearer spectrum (*156*).

Some peptidolipids already contain N-methylamino acids. Preexisting N-methyl groups and chemically introduced methyl groups can be differentiated easily by comparing the mass spectra of the permethylated and perdeuteriomethylated derivatives. For example, methylation of fortuitin (**34**) increases the molecular weight by 84 mass units which corresponds to incorporation of six methyl groups. As fortuitin contains nine amino acids but only eight NH groups because of the proline residue, two N-methyl groups are present in the initial compound. Comparison of the mass spectra of permethylated and perdeuteriomethylated fortuitin located the two N-methylamino acids (*155*).

Many peptidolipids contain a lactone ring which, it may be thought, might interfere with this regular splitting of the molecule. In fact the lactone ring is easily opened in the mass spectrometer, so the cleavage described earlier can take place. For example, the mass spectrum of peptidolipin NA (**44**) exhibits, in addition to the molecular ion at *m/z* 963, a relatively intense peak at *m/z* 919 corresponding to the loss of 44 mass units. Precise mass measurements showed that the 919 ion results from the loss of CH_3–CHO (and not CO_2), probably from the C-terminal threonine. Another less intense peak at *m/z* 883 is due to the loss of two H_2O molecules and CO_2, as shown in formula (**62**). As a consequence, the ring is opened by breaking the C-terminal amino acid so that the usual kind of fragmentation of the resulting ion can take place (*109*).

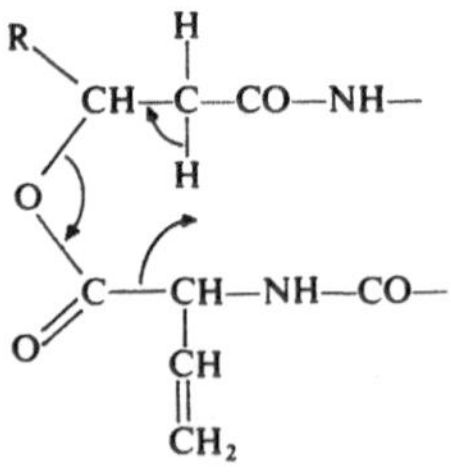

(62)

Fragmentations involving N–C cleavages have been observed in the mass spectra of peptides containing asparagine or tyrosine residues. They can be explained by the mechanism outlined in Scheme 9 (*136*) and were observed in the mass spectrum of iturin A. Tyrosine-containing peptides

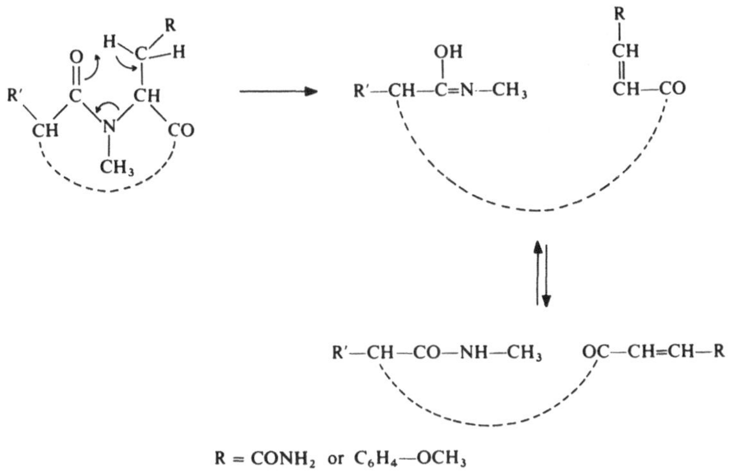

R'—CH—C(OH)=N—CH$_3$ R—CH=CH—CO

⇅

R'—CH—CO—NH—CH$_3$ OC—CH=CH—R

R = $CONH_2$ or C_6H_4—OCH_3

Scheme 9. Mechanism of N–C cleavage

give satisfactory mass spectra only after methylation of the phenolic hydroxyl group (*157*). EI mass spectrometry of permethylated peptides containing aspartic or glutamic acid residues has been studied particularly thoroughly (*158*).

No difference in the basic fragmentation pattern has been observed when a fatty acid residue is located at the N-terminal end of a peptide molecule; the length of the fatty acid was also found not to play a role (*159*).

From the $[M + H]^+$ peak, molecular weights of large peptides are easily obtained by FD mass spectrometry. Sequence-specific cleavages on both sides of the CO group of peptide bonds are observed, so the determination of the complete sequence of amino acids is difficult (*160*; see also *161*).

No special studies of peptidolipids by FAB mass spectrometry seem to have been carried out. However it has been shown that the degree of fragmentation of peptides can be increased by derivatizing them with a hydrophobic group and adding a mineral acid such as $HClO_4$ to the matrix (*162*).

6. Biological Properties of Peptidolipids

Most peptidolipids of the type discussed earlier exhibit antibiotic properties against various species of bacteria. For example, iturin A is active against *Micrococcus luteus* (*163*). They can also be active against yeasts and fungi (*164*, *165*) and even against some viruses (*121*). When a lactone ring exists in the peptidolipid, opening of the ring causes the antibiotic properties to disappear, probably because of the resulting changes in conformation. The antibiotic properties are often due to inhibition of cell wall biosynthesis; this is the case for lipopeptin A acting on bacteria (*115*) and neopeptin acting on fungi (*117*).

Many of these peptidolipids have amphiphilic properties. This is illustrated by the protoplast-bursting activity of surfactin (*93*) or by the strong lysing activity of iturin A (but not iturin C) and bacillomycin L on erythrocytes (*166*), or the disruption of the outer membrane of *Escherichia coli* by EM 49 (*152*). Likewise antibiotics of the iturin group increased the size of small unilamellar vesicles of saturated lecithins (*167*). Differences in the balance between the hydrophobicity of the hydrocarbon chain and the polarity of the peptide moiety might explain the differences of action of iturin A, mycosubtilin and bacillomycin on *Micrococcus luteus* cells and protoplasts (*168*).

As some biological properties could be explained by interactions of these peptidolipids with membrane layers (*169*, *170*), studies of these phenomena were undertaken. In particular variations of electric current through planar lipid bilayers formed by glyceromonoolein were detected when peptidolipids of the iturin group were added. Discrete current fluctuations were observed which probably arose from interactions of these peptidolipid micelles with the bilayer (*171*). The pore forming properties of these peptidolipids are dependent on the conformation of the molecule. This explains the differences in behaviour of iturin A and its derivative with a methylated hydroxyl on the D-tyrosine residue (*172*). During these studies, the conformations of iturin A and of O-methyl iturin A in pyridine solution were studied by 2D-NMR techniques. At low concentrations, these peptidolipids exhibited fluctuating structures corresponding to the formation of conducting pores; their life times were dependent on their lateral diffusion rates. At higher concentrations more stable structures were formed which produced a large increase in the ionic permeability of the lipid bilayers or rapid leakage of molecules entrapped in vesicles (*173*).

Peptidolipid A21978 C displayed antibiotic activity against Gram positive bacteria. Enzymatic hydrolysis by incubation with *Actinoplanes utahensis* resulting in removal of the fatty acid was followed by chemical attachment of another fatty acid. Of the various analogues prepared in this manner, the *n*-decanoyl analogue (daptomycin) had the least acute toxicity in the mouse and exhibited antimicrobial properties at about 0.5 μg/ml (*127*).

Acylpeptides with structures similar to that of surfactin were found to be inhibitors of cyclic adenosine-3′,5′-monophosphate phosphodiesterase. Opening of the lactone ring reduced the inhibitory activity to about one half that of the intact compound (*101*).

IV. Glycosides of Peptidolipids

So far, glycosides of peptidolipids have been found mainly in bacteria of the order Actinomycetales. Such compounds containing fatty acid, amino acid and sugar components are rather complex and the determination of their structures requires the use of sophisticated techniques.

1. Glycosides of Peptidolipids Isolated from Rhodococcus erythropolis

Rhodococcus erythropolis produces a complex mixture of peptidolipids which can be separated by chromatography into seven fractions. The three most polar ones contained glycine, L-leucine, D-*allo*isoleucine, L-threonine, L-serine, L-homoserine and D-alanine in the molar ratio 3:3:2:2:2:1:1. In two of the peptidolipids glucose was bound to the hydroxyl of a serine residue and in the third, the most strongly adsorbed, glucose was bound to the hydroxyl group of a threonine residue. All these polar compounds contained one mole of a normal-chain fatty acid and one mole of a mycolic acid residue (*74*).

2. Glucoside of N-Acylnonapeptide: Herbicolin A

In the course of a screening program for the isolation of antifungal agents a bacterial strain identified as *Erwinia herbicola* was found to produce an antibiotic fraction which was highly active against yeasts and filamentous fungi but inactive against bacteria (*174*).

By countercurrent distribution, the main component was isolated and designated as herbicolin A. It was accompanied by a minor component, herbicolin B. Both compounds had the same amino acid and fatty acid composition and differed from each other in the presence of D-glucose which was only found in herbicolin A.

Herbicolin A was labile in acidic and basic media and was inactivated by chromatography on silica gel. An intense MH^+ peak was observed in the FAB and FD mass spectra at m/z 1300. It was shifted to m/z 1302 by hydrogenation and to m/z 1318 by mild treatment with alkali.

In the mixture of amino acids obtained by hydrolysis three different forms of threonine were found, L-threonine, D-*allo*threonine and N-methyl-L-*allo*threonine. Their stereochemistry was elucidated by gas chromatography of the derivatives obtained by esterification with (+)-3-methyl-2-butanol and trimethylsilylation. Synthetic compounds were used for comparison.

Partial hydrolysis of herbicolin gave an N-acylamino acid and a mixture of di-, tri- and tetrapeptides. The N-acylamino acid and its hydrogenation product were respectively identified as N-(3-hydroxytetradecanoyl)-dehydro-α-aminobutyric acid and N-(3-hydroxy-tetradecanoyl)-α-aminobutyric acid. Identification of the various di-, tri- and tetrapeptides gave the sequence of the amino acids. The D-*allo*-threonine residue was located by comparing the behaviour, in capillary gas chromatography, of the two dipeptides D-*allo*Thr-D-Leu and L-Thr-D-Leu as trifluoroacetyl methyl esters with those of the same derivatives of a partial hydrolysate. The presence of glutamic acid in the form of glutamine was demonstrated by mass spectrometry.

The presence of arginine as C-terminal amino acid was checked by its conversion into N-methylornithine by action of diborane. Herbicolin A contains a lactone ring which can be opened by mild treatment with alkali. After chromic acid oxidation of herbicolin B and hydrolysis, both

(63) Herbicolin A

allo-threonine forms disappeared as well as the 3-hydroxytetradecanoic acid. It was concluded that the lactone ring was formed by esterification of the carboxyl group of arginine with the hydroxyl group of L-threonine. The glucose residue could not be located unambiguously but "the glucose residue is most probably linked to the hydroxyl group of 3-hydroxytetradecanoic acid" (*174*). The α-configuration of the glycosidic bond was established by ^{13}C-NMR spectroscopy.

From these results, structure (**63**) was proposed for herbicolin A. The minor component, herbicolin B, is the same compound without a glucose residue (*174*).

3. Mycosides C and Related Antigenic Compounds

Systematic examination of lipid extracts from *Mycobacterium avium* by infrared spectroscopy showed the presence of compounds exhibiting absorption bands at about 1235 cm^{-1} which was attributed to acetyl ester groups (*175*). The substances responsible for this absorption were later isolated and as a group received the name of mycoside C. They were found in mycobacteria of the MAIS group (*Mycobacterium avium, M. intracellulare, M. scrofulaceum*) and also in *M. smegmatis, M. fortuitum, M. chelonae* and some related species (*176*).

It has now been recognized that two groups of mycosides C can be distinguished, the apolar mycosides C and the antigenic polar mycosides C (*177*).

3.1. Apolar Mycosides C

Apolar mycosides C were first isolated during a search for substances responsible for an absorption band at 1235 cm^{-1} from extracts of *M. avium* (*178*), *M. scrofulaceum* (using a strain called *M. marianum* at that time) (*179*) and *M. butyricum* which was probably a strain of *M. smegmatis* (*180*). As will be shown all these compounds have the same general structure and differ only in detail.

Chemical studies performed on these mycosides showed the presence of a common tripeptide backbone D-Phe-D-*allo*Thr-D-Ala (*178, 179*) sometimes extended to the pentapeptide D-Phe-D-*allo*Thr-D-Ala-D-*allo*Thr-D-Ala (*178, 179*). However it must be kept in mind that amino acids linked as amides to fatty acids are difficult to split off by hydrolysis, so their levels are often underestimated. Moreover, when mass spectrometric measurements were performed, only the tripeptide was found

(*180–183*). The stereochemistry of the three amino acids was established by enzymatic tests (*69*).

Two different sugar molecules were found per molecule of mycoside C. These sugars were identified as 6-deoxytalose and O-methyl derivatives of rhamnose, 2,3,4-tri-O-methylrhamnose and more often 3,4-di-O-methylrhamnose (*184*). Later on (*185*) these sugars were said to be 6-deoxy-L-talose and O-methyl derivatives of L-rhamnose, but no experimental evidence for their absolute stereochemistry was provided.

Complete hydrolysis of the mycoside C isolated from *M. butyricum* (mycoside C_b) gave a lipid component which was subsequently methylated with diazomethane. In the mass spectrum of the methyl ester a molecular peak at *m/z* 466 ($C_{30}H_{58}O_3$) and peaks at *m/z* 434 (M-32) and 402 (M-[32 × 2]) were found. The elimination of two methanol units suggested the presence of a methoxy group. This was confirmed by an intense peak at *m/z* 117 corresponding to the ion $[CH_3O–CH–CH_2–COOCH_3]^+$. The methoxyl group was thus located in position 3. The ester contained one double bond as evidenced by hydrogenation. Ozonolysis of the ester gave tetracosanoic acid. Structure (**64**) was proposed for this methoxy fatty acid (*180*). It was accompanied by small amounts of homologues and of the saturated analogues; no molecular peak was observed in the mass spectra of the saturated methoxy fatty acid esters (*180*). 3D-hydroxyoctacosanoic acid was found as the lipid component of mycoside C_{1217} (*181*).

$$CH_3{-}(CH_2)_{22}{-}CH{=}CH{-}\underset{}{\overset{OCH_3}{\overset{|}{C}H}}{-}CH_2{-}COOCH_3$$

(**64**)

The same hydrolytic conditions produced not only the fatty acid discussed in the previous paragraph, but also a more complex acid. In the case of mycoside C_b, the methyl ester had m.p. 53–55 °C, $[\alpha]_D - 7°$ (chloroform). Its mass spectrum exhibited an intense peak at *m/z* 613 ($C_{39}H_{67}NO_4$) (see Scheme 10) and peaks at *m/z* 162 (**67**) and 451 (**66**). This result showed that the fatty acid was linked by an amide bond to the D-phenylalanine residue in mycoside C_b (*180*). A similar result was obtained for mycoside C_{1217} (*16*).

By partial acid hydrolysis a dipeptide was isolated and identified as D-*allo*threonyl-D-alanine. Furthermore, it was observed that mild treatment with alkali cleaved mycoside C to free 6-deoxytalose and a modified mycoside devoid of 6-deoxytalose as well as *allo*threonine. The concomitant disappearance of 6-deoxytalose and of *allo*threonine suggested β-

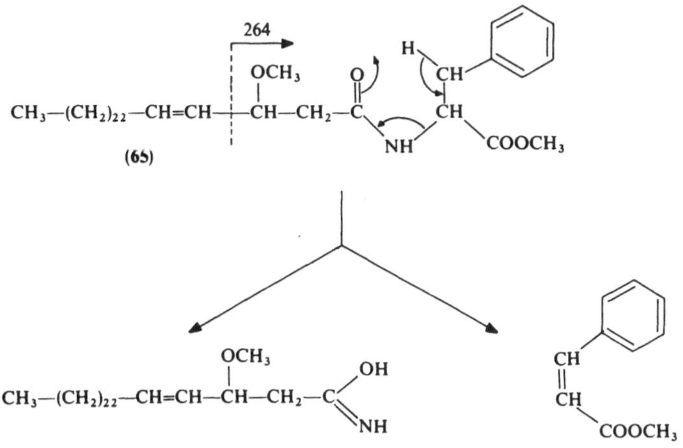

$$CH_3-(CH_2)_{22}-CH{=}CH-CH(OCH_3)-CH_2-C(=O)-NH-CH(COOCH_3)-CH_2-C_6H_5$$

(65) [264]

$$\downarrow$$

$$CH_3-(CH_2)_{22}-CH{=}CH-CH(OCH_3)-CH_2-C(OH){=}NH$$

(66) m/z 451

$$C_6H_5-CH{=}CH-COOCH_3$$

(67) m/z 162

Scheme 10

elimination of 6-deoxytalose glycosidically linked to the hydroxyl group of the *allo*threonine residue (*16*, *180*).

Mycosides C contained two acetyl groups whose IR absorption was the clue to the discovery of these molecules. They could be located on the 6-deoxytalose residue by mass spectrometry. From these results, a partial formula of mycosides C was established which contained the fatty acid moiety, the tripeptide backbone and di-O-acetyltalose linked to *allo*-threonine (*180*). However it was not clear how the second sugar, the methylated rhamnose, was linked; apparently a unit not detected in the EI mass spectra was still missing.

During the initial studies on mycoside C_{1217} the presence of small amounts of an amino alcohol, ethanolamine, had been observed (*16*). By a careful search for other similar components, the presence of alaninol was demonstrated, in particular by gas chromatography of its N,O-di-trifluoroacetyl derivative (*187*), and the first complete structure of a mycoside C (**68**) compatible with the mass spectral data, was proposed

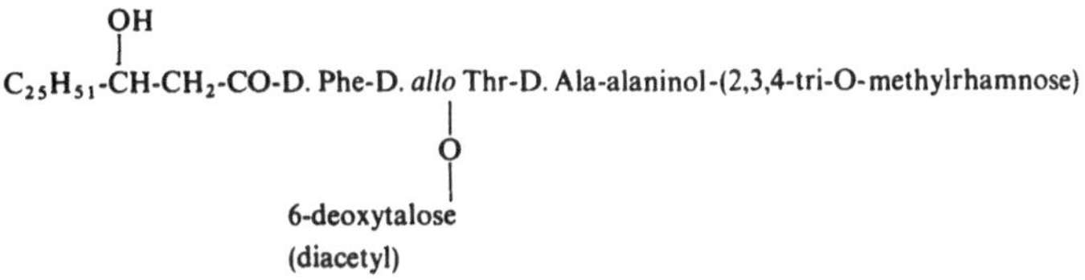

$C_{25}H_{51}$-CH(OH)-CH_2-CO-D. Phe-D. *allo* Thr-D. Ala-alaninol-(2,3,4-tri-O-methylrhamnose)

(Thr)–O–6-deoxytalose (diacetyl)

(**68**) Apolar mycoside C

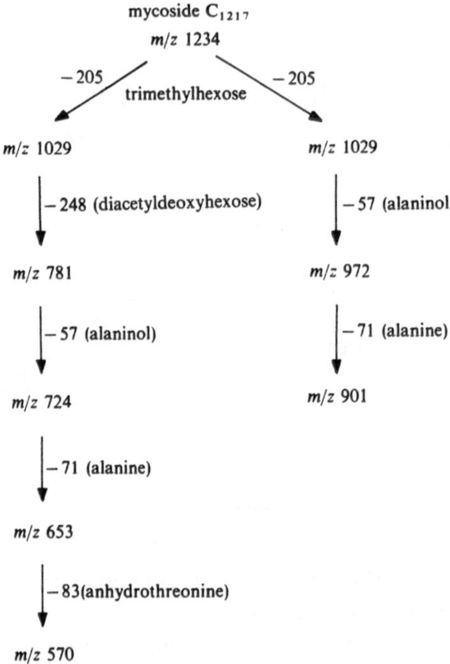

Scheme 11. Sequence of peaks in the EI-mass spectrum of mycoside C_{1217} (*188*)

(*181*). Scheme 11 shows two series of peaks observed in the mass spectrum of mycoside C_{1217} (*188*). Some molecules contained ethanolamine in the place of alaninol.

Later the complete structures of mycoside C_{b1} (*M. butyricum*) (*189*) and mycoside C_s (*M. scrofulaceum*) (*190*) were reported. A similar structure proposed for mycoside C_2 (*M. avium*) contained two puzzling features, *i.e.* the presence of palmitic acid as a fatty acid component instead of a C_{28} acid derivative and a pentapeptide backbone (*191*). N,O-dimethyl-L-serine, isolated from hydrolysates of two mycosides C (*186*) was most probably an artefact. The configuration of alaninol was determined by transformating it into alanine by chromic acid oxidation and further treatment with D-amino acid oxidase. The substance was not affected by the enzyme, showing that it was L-alaninol (*191*).

As might be expected, the D-amino acids required for the biosynthesis of mycoside C are produced by mycobacteria from the L-amino acids. An alanine reductase able to reduce L-alanine to L-alaninol was detected in a cell-free preparation of *M. avium* (*192*).

The structure of mycoside C from a strain of *M. smegmatis* was studied mainly by mass spectrometric techniques. Use of FD mass spectrometry by the cationization method gave easy access to the molecular weight. In this mycoside, the fatty acid moiety was found to be a mixture of hydroxylated and methoxylated fatty acids, mainly with 28 carbon atoms. They contained one double bond in the middle of the chain (*182*). *M. butyricum* is no longer recognized as a mycobacterial species and has been reclassified as *M. smegmatis*. In fact only minor structural differences are found between the mycosides of *M. smegmatis* and *M. butyricum* (see Table 9).

Another mycoside C devoid of 6-deoxytalose was isolated from two strains of *M. senegalense*; mono- and di-O-methylrhamnose were the only sugars obtained after hydrolysis. Of course no β-elimination was observed. Such mycosides are called mycosides C′ (*193*). The presence of glycosidases which could have hydrolyzed the glycosidic linkage of deoxytalose during collection and extraction of the bacteria cannot be excluded.

Biosynthesis

M. intracellulare (serotype 21) was grown on a medium containing [U-^{14}C]-D-alanine and was extracted with hot ethanol. Chromatographic fractionation of the extract gave three labelled glycopeptides. Structural studies showed that glycopeptide I was D-ala-L-alaninol-(3,4-di-O-Me-rhamnose), glycopeptide II, D-*allo*Thr-D-Ala-L-alaninol-(3,4-di-O-Me-rhamnose) and glycopeptide III, D-Phe-D-*allo*Thr-D-Ala-L-alaninol-(3,4-di-O-Me-rhamnose). Incubation of glycopeptide I with D-*allo*threonine, ATP and a cell-free extract of *M. intracellulare* gave rise to glycopeptide II. Addition of D-phenylalanine and ATP to the incubation mixture led to formation of glycopeptide III. No information was obtained on the transformation of glycopeptide III into mycoside C (*177*).

Distribution of mycosides C

Most studies have been performed on the mycosides C isolated from bacteria of the group MAIS. However, several other species of mycobacteria are able to synthesize mycosides C. These are *M. paratuberculosis* (*194*), *M. fortuitum* biovar *peregrinum* (*193*), *M. chelonae* (*195*), *M. senegalense* (*193*), *M. lepraemurium* (*196, 197*) and some slow-growing mycobacteria designated as ADM (Armadillo Derived Mycobacteria) (*198*).

The production of mycosides C by a given species of mycobacteria cannot be used for taxonomic purposes since mutants occur that are devoid of mycoside C. In particular it has been observed that when

Table 9. *Composition of "Apolar Mycosides C" Isolated from Various Species of Mycobacteria*

Mycobacterium		Fatty acid	O-methyl derivatives of rhamnose	Comments	References
avium	*a*	$CH_3-(CH_2)_{22}-CH=CH-CH(OCH_3)-CH_2-COOH$ $CH_3-(CH_2)_{24}-CH(OCH_3)-CH_2-COOH$	2,3,4-tri-O-methyl		(*192*)
	b	C_{16} and C_{18} fatty acids	di-O-methyl	6-deoxytalose and 3-O-methyl-6-deoxy talose	(*191*)
smegmatis		$CH_3-(CH_2)_x-CH=CH-(CH_2)_y-CH(OR)-CH_2-COOH$ R = H or CH_3 $x + y = 22$ or 24	3,4-di-O-methyl and 2,3,4-tri-O-methyl		(*182*)
butyricum		$CH_3-(CH_2)_{22}-CH=CH-CH(OCH_3)-CH_2-COOH$	2,3,4-tri-O-methyl		(*180*), (*182*)
scrofulaceum		$CH_3-(CH_2)_{25}-CH(OCH_3)-COOH$	3,4-di-O-methyl		(*190*)
marianum		?	3,4-di-O-methyl		(*179*)
sp. 1217	*c*	$CH_3-(CH_2)_{24}CH(OH)-CH_2-COOH$	3,4-di-O-methyl		(*181*)
	d	–id–	2,3,4-tri-O-methyl	ethanolamine instead of alaninol	(*16*), (*188*)
senegalense		$CH_3-(CH_2)_{24}-CH(OCH_3)-CH_2-COOH$ or 22	3-O-methyl and 3,4-di-O-methyl	no 6-deoxytalose	(*193*)

(*M. scrofulaceum* and *M. marianum* are identical species; most of *M. butyricum* strains are strains of *M. smegmatis*).
a and *b*: two different preparations; *c*: purified preparation; *d*: minor component.

bacteria of the MAIS group were grown as pellicles on the surface of the medium rough variants devoid of mycoside C were obtained in high frequency (*199*).

3.2. Polar Mycosides C: Antigenic Glycopeptidolipids

Because of the increase in tuberculosis-like diseases due to mycobacteria other than tubercle bacilli, attention has been given to easy means of characterization of such bacteria. SCHAEFER observed that most non-tuberculous mycobacteria were endowed with highly immunogenic species- or type-specific antigens (*200*). On this basis he devised a seroagglutination assay that allowed recognition of at least 31 antigenically distinct serotypes within the MAIS complex (*201*). Using different methodology a chromatographic procedure for studying whole-lipid extracts of bacteria derived from the MAIS serotypes allowed identification of the various subspecies (*202*); this procedure was based mainly on the occurrence of spots of glycolipids on thin-layer chromatograms. BRENNAN and GOREN (*183*) demonstrated that the type-specific antigens of SCHAEFER (*201*) and the glycolipids observed on thin-layer

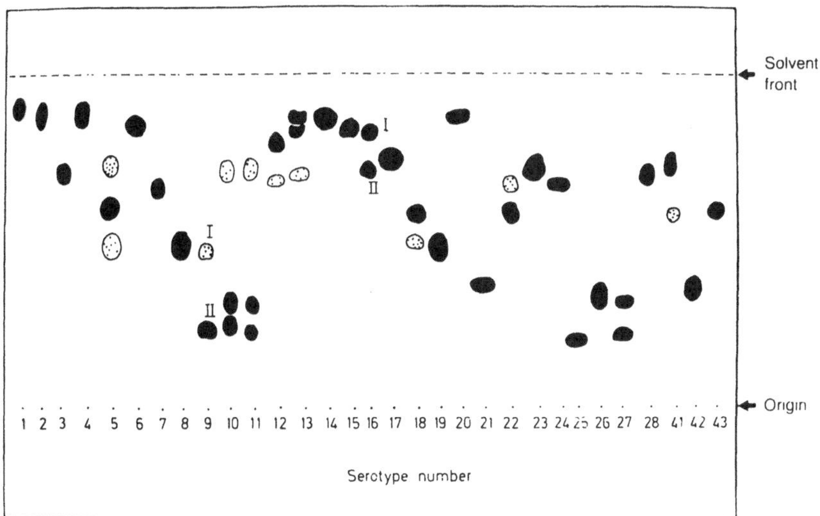

Fig. 5. Thin-layer chromatogram of the deacylated polar mycosides C from the 31 serovars of the MAIS complex. (Silicagel plate, developed with $CHCl_3$–CH_3OH–H_2O 65:25:4 |v/v/v| and detected by orcinol reagent.) (From *204*)

chromatograms by JENKINS (*202*) were the same glycopeptidolipid derivatives of mycosides C. Methods were devised to characterize these glycopeptidolipids (*203*) and to determine their structures (*204*). Figure 5 shows the thin-layer chromatographic behaviour of the 31 type-specific antigens.

Chemical studies performed on some of these antigens demonstrated that they were apolar mycosides C with 3,4-di-O-methylrhamnose and bearing an oligosaccharide chain linked to the residue of 6-deoxytalose (*183*). Table 10 reports the instances in which the structures of the oligosaccharide chains have been established. The general approach used in these structure determinations consisted of the release of the oligosaccharide chain by base-catalyzed reductive β-elimination followed by a combination of gas chromatography/mass spectrometry, methylation analysis, FAB mass spectrometry and ^{13}C- and ^{1}H-nuclear magnetic resonance spectroscopy. In some cases gas chromatography of the trimethylsilyl derivatives of the (+)-2-butyl glycosides was used to establish the absolute configurations of the sugars after demethylation (*212*). The structures of the oligosaccharide chains of the polar mycosides C serovar 9 and 25 are given as examples in Fig. 6. It can be noticed that the structures of the antigenic oligosaccharide chains of MAIS serovar 2 and *M. paratuberculosis* were the same; thus "a close relationship, if not complete homology, exists between *M. avium* serotype 2 and the causative agent of at least some forms of paratuberculosis" (*205*).

The polar mycosides C isolated from mycobacteria serovar 14 and 25 contained an unusual amino sugar which had been isolated previously from the lipooligosaccharides of *M. kansasii* and called kansosamine. Its

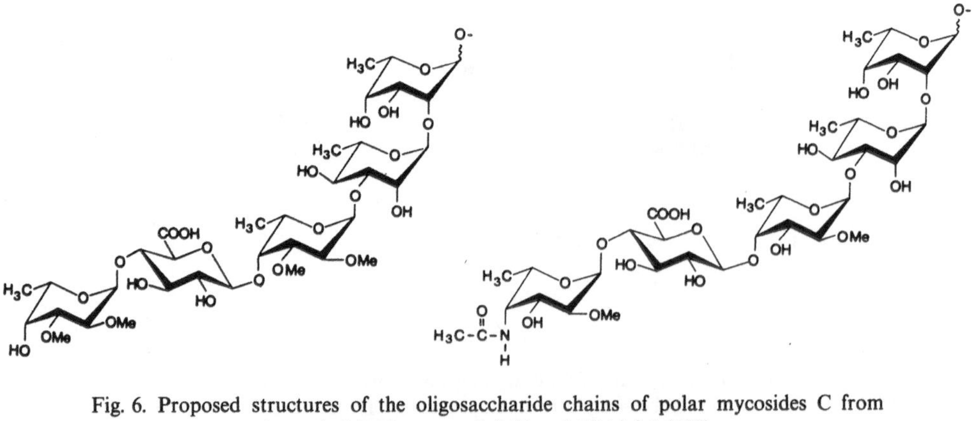

Fig. 6. Proposed structures of the oligosaccharide chains of polar mycosides C from bacteria MAIS serovar 9 (left) and 25 (right) (*207*)

Table 10. *Structures of the Oligosaccharidic Chains Linked Through the Invariant Disaccharide Rhamnosyl- 6-deoxytalose to the Hydroxyl of* D-*allothreonine in the Polar Mycosides C*

MAIS serovars	Oligosaccharidic chains, linked to hydroxyl 3 of L-Rha*p*-(α1-2)-6-deoxy-L-talose	References
2	2,3-di-O-Me L-Fuc*p*-(α1-	(*205*)
4	4-O-Me L-Rha*p*-(α1-4)-2-O-Me L-Fuc*p*-(α1-	(*206*)
8	4,6-(1′-carboxyethylidene)-3-O-Me D-Glc*p*-(β1-	(*207*)
9	2,3-di-O-Me L-Fuc*p*-(α1-4)-D-GlcA*p*-(β1-4)-2,3-di-O-Me L-Fuc*p*-(α1-	(*183, 207*)
12	4-(2′hydroxy)propionamido-4,6-dideoxy-3-OMe D-Glc*p*-(β1-3)-4-O-Me L-Rha*p*-(α1-3)-L-Rha*p*-(α1-	(*208*)
14	N-formylkansosaminyl-(1-3)-2-O-Me D-Rha*p*-(α1-3)-2-O-Me Fuc*p*-(α1-	(*209*)
19	3,4-di-O-Me D-GlcA*p*-(β1-3)-2,4-di-O-Me 3-C-Me, 3,6-dideoxyhexosyl-(1-3)-L-Rha*p*-(α1-	(*210*)
20	2-O-Me Rha-2-O-Me Fuc -(α1-	(*207*)
25	4-acetamido-4,6-dideoxy 2-O-Me hexosyl-(α1-4)-D-GlcA*p*-(β1-4)-2-O-Me L-Fuc*p*-(α1-	(*211*)
M. paratuberculosis	2,3-di-O-Me L-Fuc*p*-(α1-	(*194*)

p = pyranose; GlcA = glucuronic acid.

structure was established as 4,6-dideoxy-2-O-methyl-3-C-methyl-4-amino-L-mannose (**69**), with a branched carbon chain (*213*). Another branched chain sugar (**70**) was found in the polar mycoside C of mycobacteria serotype 19 (see Table 10).

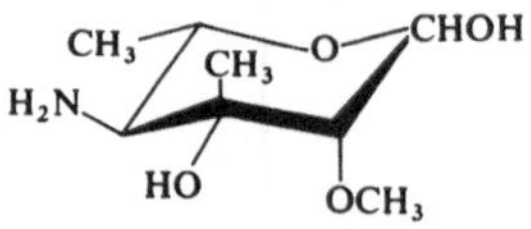

(**69**) 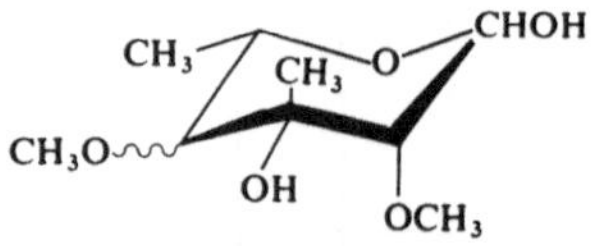 (**70**)

As the polar mycosides C behave as antigens they must be located in the outer layers of the mycobacterial cell. In fact their external location was demonstrated in *M. intracellulare* by the use of ferritin-labelled antibodies (*214*). It has long been known that mycobacteria of the MAIS complex are surrounded by an "electron-transparent zone" when ultra-thin sections of infected host tissues are examined under the electron microscope. This zone is essentially composed of parallel fibrils wrapped longitudinally around the bacterium. This material was isolated by the use of urea density gradients and analyzed chemically. The three amino acids characteristic of mycosides C and alaninol as well as 6-deoxytalose were identified (*194*). This material consisted mainly of a mixture of polar and apolar mycosides C, in a ratio about 7:3 (*215*).

Polar mycosides C from serovars 4 and 20 were prepared in radio-active form by growing suitable strains of *M. avium* on media containing [6-^{3}H]-fucose, [2-^{3}H]-mannose or [methyl-^{3}H]-methionine, in order to study their behaviour inside macrophages (*216*).

It was observed that *o*-, *p*- and *m*-monofluoro-DL-phenylalanine caused 80–90% inhibition of mycoside C biosynthesis in a strain of *M. avium*. Ultrastructural observation of the bacteria treated with the *meta* isomer showed profound alterations in the outer wall layer (*217*). In the course of a study on the multiplication and intracellular survival of *M. avium* inside macrophages, comparisons of smooth and rough variants (both producing mycosides C) and a mutant unable to synthesize mycoside C led to the conclusion that the mycosides C were not primary factors involved in the survival of mycobacteria inside macrophages (*218*).

3.3. Mycosides C as Receptors for Mycobacteriophages

Mycobacteriophage D4 is a phage specific for *M. smegmatis* and its adsorption onto cells of *M. smegmatis* can be inhibited by compounds that behave as receptors for this phage. This inhibition of adsorption of phage D4 by apolar mycosides C was shown by two groups (*219*, *220*), the inhibition being independent of the origin of the mycoside C (*M. smegmatis*, *M. scrofulaceum* or *M.* sp. 1217). No adsorption occurred when mycobacteriophage D29 was used instead of D4.

Under assay conditions where apolar mycosides C produced 85–90% inhibition of the lysis of *M. smegmatis* by a given dose of mycobacteriophage D4 deacetylated mycoside C gave only 10–20% inhibition, whereas the unsaturated product arising from β-elimination of deoxytalose gave about 50% inhibition (*221*). Thus deacetylation of the di-O-acetyl-6-deoxytalose moiety dramatically decreased the ability of the mycoside C to adsorb phage D4, while its complete disappearance had much less effect. An explanation was proposed based on the hydrogen bonding abilities of the freed hydroxyl groups. "Although the acetylated talose may play an auxiliary role in D4 adsorption, it does not itself act as the receptor" (*221*).

Fifty five to sixty per cent inhibition was exhibited by compounds as simple as octadecyl triacetyl-L-rhamnoside or octadecyl tri-O-methyl-rhamnoside (probably a mixture of anomers), whereas similar derivatives of D-fucose and 6-deoxy-L-talose had no inhibitory properties, showing the importance of the substituted rhamnose residue in this phenomenon. It must be underlined that octadecyl L-rhamnoside with free hydroxyl groups had only little activity (*221*).

Polar mycosides C isolated from various serovars exhibited considerably less adsorptive activity for mycobacteriophage D4 than did the apolar mycosides. It was proposed that the oligosaccharide chain of the polar mycosides hindered the site of the molecule that was recognized by the phage. As the mycobacteria which synthesize the polar mycosides C also produce apolar mycosides and as they were not sensitive to phage D4, it was assumed that the polar mycosides C were located in a more external layer than the apolar ones and prevented contact between the tails of the phages and the recognition site of mycoside C (*221*).

V. Conclusion

Peptidolipids seem to be located mainly at the surface of the bacterial cell or may even be excreted into the surrounding medium for the more hydrophilic members. Because of their amphiphilic character and their location they are able to interfere with many vital processes in surrounding cells. Thus many peptidolipids exhibit hemolytic properties, antibiotic activities – often against fungi – or enzyme-inhibiting properties.

Most of these peptidolipids contain one or several D-amino acids and some unusual amino acids such as *threo*-3-hydroxy-L-glutamic acid and cysteic acid. The peptide chain of imacidin C includes D-*allo*threonine and the much rarer L-*allo*threonine. The presence of such amino acids probably plays a role in increasing the resistance of these substances towards proteolytic enzymes. For instance, imacidic acid or imacidin is not cleaved by any of eight different proteolytic enzymes (trypsin, chymotrypsin, pepsin, thermolysine, pronase, carboxypeptidases A and B, papain) (*124*). This could be very useful for protection of the bacterial cell or for conservation of the molecule inside the host tissues.

So far little work has been devoted to the metabolism of peptidolipids or, except in a few instances, to their biosynthesis. Their physico-chemical and biological properties also need further investigation.

The D-amino acid residues also play a role in the conformation of the molecules. The properties of a peptidolipid such as, for example, the antibiotic activities, can be preserved after an exchange of amino acids provided that the D- or L-structures remain in the right place (see for instance Table 7).

Many peptidolipids contain a macrocycle of the lactone or lactam type. The size of this cycle varies depending on the peptidolipid as it can be formed from 5 to 10 amino acids. However the most frequently encountered lactone or lactam rings arise from cyclization of a chain of seven amino acid residues, thus producing cycles with 22 to 25 atoms. The presence of a ring seems to stabilize preferred conformations and the simple opening of the lactone ring causes the antibiotic properties to disappear. Surfactin and esperin are unique in that the cyclization of the same N-acyl heptapeptide has produced a ring incorporating all seven amino acids to give surfactin (**40**) or a ring incorporating only five aminoacids to give esperin (**43**). It is a pity that no work seems to have been done to investigate the surface active properties of esperin compared with those of surfactin.

The discovery of new members of this group of bacterial metabolites is often a matter of chance, although systematic screening aimed at isolation of antibiotics or enzyme inhibitors can play a role. New peptidolipids, some of whose structures have not yet been determined, continue to be isolated. Most likely this group of natural products will continue to grow in quantity and importance in the near future and there is little doubt that the role that they play in the life of the bacteria which produce them and the applications that they may have will be investigated more intensively.

Acknowledgements

Many helpful discussions with Dr. C. ASSELINEAU are gratefully acknowledged. Warm thanks are expressed to Mrs. J. LECOINTE for her ability to find rare documents.

Addendum

After completion of this review, two papers were published dealing with structure determination of new peptidolipids, each of which contains chlorine.

The cyanobacteria *Anabaena* BQ 16-1 produces a decapeptide linked to a β-amino fatty acid in the same manner as in the iturins. The β-amino fatty acid was identified as 2-hydroxy-3-amino-4-methyl-14-chloropalmitic acid. This peptidolipid designated as *puwainaphycine C* exhibits cardiotonic activity. It is accompanied by puwainaphycine *D* which differs from the C-compound in that a threonine residue is replaced by a valine residue. This slight modification causes disappearance of the cardiotonic activity (*222*).

The structure of the phytotoxin produced by *Pseudomonas syringae* pathovar *syringae* has been elucidated (*223*). It is an N-acylpeptolide named syringomycin E, with a lactone ring made up of nine amino acids. In the N-terminal amino acid, serine, the hydroxyl group is esterified by the β-carboxyl group of the C-terminal amino acid, aspartic acid, and the amino group is acylated by 3-hydroxydodecanoic acid. The chlorine atom is found in the eighth amino acid residue which is 4-chloro-threonine (*223*).

Both of these peptidolipids on hydrolysis unexpectedly lose the chlorine atom. "No explanation can be given at this time for the facile loss of the chloro group from C-14 during acid hydrolysis" of puwainaphycin D (*222*).

References

1. HOUTSMULLER, U.M.T., and L.L.M. VAN DEENEN: On the Accumulation of Amino Acid Derivatives of Phosphatidylglycerol in Bacteria. Biochim. Biophys. Acta **84**, 96 (1964).

2. LEROUGE, P., M.-H. LEBAS, C. AGAPAKIS-CAUSSE, and J.-C. PROME: Isolation and Structural Characterization of a New Non-phosphorylated Lipoamino Acid from *Mycobacterium phlei*. Chem. Phys. Lipids **49**, 161 (1988).

3. KATES, M.: Techniques of Lipidology, 2nd edition, Elsevier, Amsterdam, p. 109 (1986).

4. WELLS, M.A., and J.C. DITTMER: The Use of Sephadex for the Removal of Non-lipid Contaminants from Lipid Extracts. Biochemistry **2**, 1259 (1963).

5. WILLIAMS, J.P., and P.A. MERRILEES: The Removal of Water and Non-lipid Contaminants from Lipid Extracts. Lipids **5**, 367 (1969).
6. CARTWRIGHT, N.J.: Serratamic acid, a Derivative of L-Serine Produced by Organisms of the *Serratia* group. Biochem. J. **60**, 238 (1955).
7. CARTWRIGHT, N.J.: The Structure of Serratamic acid. Biochem. J. **67**, 663 (1957).
8. WASSERMAN, H.H., J.J. KEGGI, and J.E. MCKEON: Serratamolide, A Metabolic Product of *Serratia*. J. Am. Chem. Soc. **83**, 4107 (1961).
9. WASSERMAN, H.H., J.J. KEGGI, and J.E. MCKEON: The Structure of Serratamolide. J. Am. Chem. Soc. **84**, 2978 (1962).
10. SHEMYAKIN, M.M., YU. OVCHINNIKOV, V.K. ANTONOV, A.A. KIRYUSHKIN, V.T. IVANOV, V.I. SHCHELOKOV, and A.M. SHKROB: Total Synthesis of Serratamolide. I. Synthesis of O,O′-Diacetylserratamolide. Tetrahedron Letters 47 (1964).
11. BERMINGHAM, M.A.C., B.S. DEOL, and J.L. STILL: The Occurrence of Bound Serine in Acetone Extracts of *Serratia marcescens* Exclusively in Compounds of a Cyclic Depsipeptide Structure Related to Serratamolide. Biochem. J. **116**, 759 (1970).
12. BERMINGHAM, M.A.C., B.S. DEOL, and J.L. STILL: The Relationship Between Prodigiosin Biosynthesis and Cyclic Depsipeptides in *Serratia marcescens*. J. Gen. Microbiol. **67**, 319 (1971).
13. KAWAI, Y., I. YANO, and K. KANEDA: Various Kinds of Lipoamino Acids Including a Novel Serine-containing Lipid in an Opportunistic Pathogen *Flavobacterium*. Their Structures and Biological Activities on Erythrocytes. Eur. J. Biochem. **171**, 73 (1988).
14. GENDRE, T., and E. LEDERER: Sur les substances azotées des phosphatides de quelques mycobactéries. Ann. Acad. Sci. Fennicae, Ser. A II, **60**, 313 (1955).
15. LANEELLE, M.A., G. LANEELLE, and J. ASSELINEAU: Sur la présence d'ornithine dans des lipides bactériens. Biochim. Biophys. Acta **70**, 99 (1963).
16. LANEELLE, M.A., G. LANEELLE, P. BENNET, and J. ASSELINEAU: Sur les lipides d'une souche non-photochromogène de mycobactérie. Bull. Soc. chim. biol. **47**, 2047 (1965).
17. GORCHEIN, A.: Ornithine in *Rhodopseudomonas sphaeroides*. Biochim. Biophys. Acta **84**, 356 (1964).
18. GORCHEIN, A.: Studies on the Structure of an Ornithine-Containing Lipid from Nonsulfur Purple bacteria. Biochim. Biophys. Acta **152**, 358 (1968).
19. GORCHEIN, A.: Structure of the Ornithine-Containing Lipid from *Rhodopseudomonas sphaeroides*. Biochim. Biophys. Acta **306**, 137 (1973).
20. GORCHEIN, A.: Distribution and Metabolism of Ornithine in *Rhodopseudomonas sphaeroides*. Proc. Roy. Soc., Ser. B, **170**, 265 (1968).
21. MARINETTI, G.V., and K. CATTIEU: Lipid Analysis of Cells and Chromatophores of *Rhodopseudomonas sphaeroides*. Chem. Phys. Lipids **28**, 241 (1981).
22. RIVAS, E.A., N.L. KERBER, A.A. VIALE, and A.F. GARCIA: Isolation of a "Basic Membrane" Fraction Enriched in an Ornithine-Containing Lipid, from a Blue-green Mutant of *Rhodospirillum rubrum*. FEBS-Letters **11**, 37 (1970).
23. BROOKS, J.L., and A.A. BENSON: Studies on the Structure of an Ornithine-Containing Lipid from *Rhodospirillum rubrum*. Arch. Biochem. Biophys. **152**, 347 (1972).
24. THIELE, O.W., J. OULEVEY, and D.H. HUNEMAN: Ornithine-Containing Lipids in *Thiobacillus* A2 and *Achromobacter* Sp. Eur. J. Biochem. **139**, 131 (1984).
25. GHOSH, M., and A.K. MISHRA: Occurrence, Identification and Possible Significance of Ornithine Lipid in *Thiobacillus ferrooxidans*. Biochem. Biophys. Res. Comm. **142**, 925 (1987).
26. KNOCHE, H.W., and J.M. SHIVELY: The structure of an Ornithine-Containing Lipid from *Thiobacillus thiooxidans*. J. Biol. Chem. **247**, 170 (1972).

27. DEES, C., and J. M. SHIVELY: Localization and Quantitation of the Ornithine Lipid of *Thiobacillus thiooxidans*. J. Bacteriol. **149**, 798 (1982).
28. WILKINSON, S.G.: Cell Walls of *Pseudomonas* Species Sensitive to Ethylenediamine Tetraacetic acid. J. Bacteriol. **104**, 1035 (1970).
29. MINNIKIN, D.E., and H. ABDOLRAHIMZADEH: The Replacement of Phosphatidylethanolamine and Acidic Phospholipids by an Ornithine-Amide Lipid and Minor Phosphorus-free Lipids in *Pseudomonas fluorescens* NCMB 129. FEBS-Letters **43**, 257 (1974).
30. PITTA, T.P., E.R. LEADBETTER, and W. GODCHAUX: Increase of Ornithine Amino Lipid Content in a Sulfonolipid-Deficient Mutant of *Cytophaga johnsonae*. J. Bacteriol. **171**, 952 (1989).
31. PROME, J.-C., C. LACAVE, and M.-A. LANEELLE: Sur les structures de lipides à ornithine de *Brucella melitensis* et de *Mycobacterium bovis* (BCG). Compt. Rend. Acad. Sci. (Paris) Ser. C, **269**, 1664 (1969).
32. TAHARA, Y., Y. YAMADA, and K. KONDO: A New Lysine-Containing Lipid Isolated from *Agrobacterium tumefaciens*. Agr. Biol. Chem. **40**, 1449 (1976).
33. KIMURA, A., and H. OTSUKA: The Changes of Lysine- and Ornithine-Lipids in *Streptomyces sioyaensis*. Agr. Biol. Chem. **33**, 781 (1969).
34. THIELE, O.W., and G. SCHWINN: The Free Lipids of *Brucella melitensis* and *Bordetella pertussis*. Eur. J. Biochem. **34**, 333 (1973).
35. THIELE, O.W., and G. SCHWINN: Bakterielle Ornithin-Lipid. Zeits. allg. Mikrobiol. **14**, 435 (1974).
36. KAWANAMI, J., A. KIMURA, and H. OTSUKA: Siolipin, a New Lipoamino Acid Ester Isolated from *Streptomyces sioyaensis*. Biochim. Biophys. Acta **152**, 808 (1968).
37. KAWANAMI, J.: Lipids of *Streptomyces toyocaensis*. On the Structure of Siolipin. Chem. Phys. Lipids **7**, 159 (1971).
38. LANEELLE, M.A., G. LANEELLE, D. PROME, and J.C. PROME: Ornithine Lipids of *Mycobacterium tuberculosis* and Some Other Mycobacteria. J. Gen. Microbiol., **136**, 773 (1990).
39. IMHOFF, J., D.J. KUSHNER, S.C. KUSHWAHA, and M. KATES: Polar Lipids in Phototrophic Bacteria of the *Rhodospirillaceae* and *Chromataceae* Families. J. Bacteriol. **150**, 1192 (1982).
40. BATRAKOV, S.G., and L.D. BERGELSON: Lipids of the Streptomycetes. Structural Investigation and Biological Interrelation. Chem. Phys. Lipids **21**, 1 (1978).
41. WEE, S., and B. J. WILKINSON: Increased Outer Membrane Ornithine-Containing Lipid and Lysozyme Penetrability of *Paracoccus denitrificans* Grown in a Complex Medium Deficient in Divalent Cations. J. Bacteriol. **170**, 3283 (1988).
42. TAHARA, Y., K. SHINMOTO, Y. YAMADA, and K. KONDO: A New Enzyme, Acyl-CoA: N^{α}-3-hydroxypalmitoyl-L-Ornithine O-acyl Transferase. Agr. Biol. Chem. **42**, 1447 (1978).
43. FUKUDA, H., S. IWADE, and A. KIMURA: A New Enzyme: Long Acyl Aminoacylase from *Pseudomonas diminuta*. J. Biochem. (Tokyo) **91**, 1731 (1982).
44. SHINTANI, Y., H. FUKUDA, N. OKAMOTO, K. MURATA, and A. KIMURA: Isolation and Characterization of N-Long Chain Acyl Aminoacylase from *Pseudomonas diminuta*. J. Biochem. (Tokyo) **96**, 637 (1984).
45. TAHARA, Y., M. KAMEDA, Y. YAMADA, and K. KONDO: A New Lipid, the Ornithine and Taurine-Containing "Cerilipin". Agr. Biol. Chem. **40**, 243 (1976).
46. TAHARA, Y., K. SHINMOTO, Y. YAMADA, and K. KONDO: Enzymatic Synthesis of Tauro-ornithine Lipid in *Gluconobacter cerinus*. Agr. Biol. Chem. **42**, 205 (1978).

47. HILKER, D.R., M.L. GROSS, H.W. KNOCKE, and J.M. SHIVELY: The Interpretation of the Mass Spectrum of an Ornithine-Containing Lipid from *Thiobacillus thiooxidans*. Biomed. Mass Spectrom. **5**, 64 (1978).
48. HILKER, D.R., H.W. KNOCKE, and M.L. GROSS: Thermolysis Chemical Ionization of Complex Polar Lipid. Biomed. Mass Spectrom. **6**, 356 (1979).
49. MADHAVAN, V.N., J. DONE, and J. VINE: Characterization of Two Ornithine-Containing Lipids from *Erwinia aroideae*. Chem. Phys. Lipids **28**, 79 (1981).
50. ASSELINEAU, J., F. PICHINOTY, D. PROME, and J.-C. PROME: Composition des lipides complexes de *Flavobacterium meningosepticum*. Ann. Inst. Pasteur/Microbiol. **139**, 159 (1988).
51. GROSS, M.L.: Triple Sector Instrument for Tandem Mass Spectrometry. In: Tandem Mass Spectrometry, ed. by F.W. MCLAFFERTY, John Wiley & Sons Ltd., Chichester, 255 (1983).
52. TOMER, K.B., F.W. CROW, H.W. KNOCKE, and M.L. GROSS: Fast Atom Bombardment and Mass Spectrometry/Mass Spectrometry for Analysis of Ornithine-Containing Lipids from *Thiobacillus thiooxidans*. Analyt. Chemistry **55**, 1033 (1983).
53. KIMURA, A., and H. OTSUKA: Biological Activities of Siolipin (Ester of Lipoamino Acid). Agr. Biol. Chem. **33**, 1291 (1969).
54. KAWAI, Y., A. MORIBAYASHI, and I. YANO: Ornithine-Containing Lipid of *Bordetella pertussis*. J. Bacteriol. **152**, 907 (1982).
55. KAWAI, Y., and I. YANO: Ornithine-Containing Lipid of *Bordetella pertussis*, a New Type of Hemagglutinin. Eur. J. Biochem. **136**, 531 (1983).
56. KAWAI, Y., and A. MORIBAYASHI: Characteristic Lipids of *Bordetella pertussis*: Simple Fatty Acid Composition, Hydroxy Fatty Acids and an Ornithine-Containing Lipid. J. Bacteriol. **151**, 996 (1982).
57. KAWAI, Y., K. SUZUKI, and T. HAGIWARA: Phosphatidylserine and Ornithine-Containing Lipids of *Bordetella*, Hemagglutinins of Lipoamino Structure and their Control in Biomembranes. Eur. J. Biochem. **147**, 367 (1985).
58. HOLT, S.C., J. DOUNDOWLAKIS, and B.J. TABAKS: Phospholipid Composition of Gliding Bacteria: Oral Isolates of *Capnocytophaga* Compared with *Sporocytophaga*. Infect. and Immun. **26**, 305 (1979).
59. MAKULA, R.A., and W.R. FINNERTY: Isolation and Characterization of an Ornithine-Containing Lipid from *Desulfovibrio gigas*. J. Bacteriol. **123**, 523 (1975).
60. TAHARA, Y., Y. YAMADA, and K. KONDO: Occurrence of Phosphatidylcholine in *Gluconobacter cerinus*. Agr. Biol. Chem. **39**, 2261 (1975).
61. TAHARA, Y., M. KAMEDA, Y. YAMADA, and K. KONDO: An Ornithine-Containing Lipid Isolated from *Gluconobacter cerinus*. Biochim. Biophys. Acta **450**, 225 (1976).
62. THIELE, O., C.J. BISWAS, and D.H. HUNEMAN: Isolation and Characterization of an Ornithine-Containing Lipid from *Paracoccus denitrificans*. Eur. J. Biochem. **105**, 267 (1980).
63. KAWAI, Y., I. YANO, K. KANEDA, and E. YABUUCHI: Ornithine-Containing Lipids of some *Pseudomonas* Species. Eur. J. Biochem. **175**, 633 (1988).
64. COX, A.D., and S.G. WILKINSON: Polar Lipids and Fatty Acids of *Pseudomonas cepacia*. Biochim. Biophys. Acta **1001**, 60 (1989).
65. WILKINSON, S.G.: An Ornithine-Containing Lipid from *Pseudomonas rubescens*. J. gen. Microbiol. **68**, vi (1971).
66. KIMURA, A., J. KAWANAMI, and H. OTSUKA: Lipids of *Streptomyces sioyaensis*. J. Biochem. (Tokyo) **62**, 384 (1967).
67. KAWANAMI, J., and H. OTSUKA: Lipids of *Streptomyces sioyaensis*. VI. On the β-Hydroxy Acids in Siolipin. Chem. Phys. Lipids **3**, 135 (1969).

68. KIMURA, A., J. KAWANAMI, and H. OTSUKA: Distribution of Siolipin in Comparison with Phospholipids. Agric. Biol. Chem. **33**, 790 (1969).
69. IKAWA, M., E.E. SNELL, and E. LEDERER: Occurrence of D-Phenylalanine, D-*Allo*-threonine and other D-Amino acids in Peptidolipids of Bacterial Origin. Nature **188**, 558 (1960).
70. IKAWA, M., and E.E. SNELL: The Occurrence of D-*Allo*isoleucine and D-Leucine in Mycosides and Peptidolipids of Bacterial Origin. Biochim. Biophys. Acta **60**, 186 (1962).
71. MARNER, F.J., R.E. MOORE, K. HIROTSU, and J. CLARDY: Majusculamides A and B, Two Epimeric Lipodipeptides from *Lyngbya majuscula* Gomont. J. Org. Chem. **42**, 2815 (1977).
72. BATRAKOV, S.G., V.B. MURATOV, L.D. BERGELSON, and T.V. KORONELLI: Lipids of Mycobacteria. VI. Peptidolipids of the Paraffin-Oxidizing Bacterium *Mycobacterium paraffinicum*. Isolation and General Characterization. Bioorgan. Khim. **7**, 1075 (1981).
73. BATRAKOV, S.G., V.B. MURATOV, B.V. ROZYNOV, L.D. BERGELSON, and T.V. KORONELLI: Lipids of Mycobacteria. VII. N-Acyl Tetrapeptide with a Mycolic Acid Residue from *Mycobacterium paraffinicum*. Bioorgan. Khim. **7**, 1087 (1981).
74. KORONELLI, T.V.: Investigation of the Lipids of Saprophytic Mycobacteria in the USSR. J. Chromatog. **440**, 479 (1988).
75. LANEELLE, G. and J. ASSELINEAU: Isolement de peptidolipides à partir de *Mycobacterium paratuberculosis*. Biochim. Biophys. Acta **59**, 731 (1962).
76. LANEELLE, G., J. ASSELINEAU, W.A. WOLSTENHOLME, and E. LEDERER: Détermination de séquences d'acides aminés dans des oligopeptides par la spectrométrie de masse. III. Structure d'un peptidolipide de *Mycobacterium johnei*. Bull. Soc. Chim. Fr. 2133 (1965).
77. LANEELLE, G.: Etude de deux formes lipophiles d'acides aminés produites par des Mycobactéries. Thèse Doctorat-es-Sciences, Université de Toulouse (1967).
78. LANEELLE, G.: Mise en évidence d'une conformation stable d'un peptidolipide. FEBS-Letters **4**, 210 (1969).
79. HASENBOEHLER, A., H. KNEIFEL, W.A. KOENIG, H. ZÄHNER, and H.J. ZEILER: Metabolites of microorganisms. 134. Stenothricin, a New Inhibitor of Bacterial Cell Wall Synthesis. Arch. Microbiol. **99**, 307 (1974).
80. KOENIG, W.A., C. ENGELFRIED, H. HAGENMAIER, and H. KNEIFEL: Struktur des Peptidantibiotikums Stenothricin. Liebigs Ann. Chem. 2011 (1976).
81. VILKAS, E., A.M. MIQUEL, and E. LEDERER: Sur l'isolement et la structure de la fortuitine, peptidolipide de *Mycobacterium fortuitum*. Biochim. Biophys. Acta **70**, 217 (1963).
82. BARBER, M., P. JOLLES, E. VILKAS, and E. LEDERER: Determination of Amino acid Sequences in Oligopeptides by Mass Spectrometry. I. The Structure of Fortuitin, an Acylnonapeptide Methyl Ester. Biochem. Biophys. Res. Comm. **18**, 469 (1965).
83. PTAK, M., E. VILKAS, and C. BREVARD: A 400 MHz ^{1}H NMR Study of Fortuitin, a Natural Linear Lipopeptide. Biochem. Biophys. Res. Comm. **113**, 121 (1983).
84. SHOJI, J., and T. KATO: The Amino Acid Sequence of Cerexin A. J. Antibiotics **28**, 764 (1975).
85. SHOJI, J., T. KATO, and R. SAKAZAKI: The Total Structure of Cerexin A. J. Antibiotics **29**, 1268 (1976).
86. RESSLER, C., and D.V. KASHELIKAR: Identification of Asparaginyl and Glutaminyl Residues in *endo* Position in Peptides by Dehydration-Reduction. J. Am. Chem. Soc. **88**, 2025 (1966).
87. KIMURA, Y., and N. YASUDA: Polymyxin Acylase: Purification and Characterization

with Special Reference to Broad Substrate Specificity. Agric. Biol. Chem. **53**, 497 (1989).

88. SHOJI, J., T. KATO, S. TERABE, and R. KONAKA: Resolution of Peptide Antibiotics Cerexins and Tridecapeptins by High Performance Liquid Chromatography. J. Antibiotics **32**, 313 (1979).
89. SHOJI, J., H. HINOO, R. SAKAZAKI, T. KATO, Y. WAKISAKA, M. MAYAMA, S. MATSUURA, and H. MIWA: Isolation of Tridecapeptins A, B and C. J. Antibiotics **31**, 646 (1978).
90. KATO, T., R. SAKAZAKI, H. HINOO, and J. SHOJI: The Structures of Tridecapeptins B and C. J. Antibiotics **32**, 305 (1979).
91. BODANSZKY, M., G.F. SIGLER, and A. BODANSZKY: Structure of the Peptide Antibiotic Amphomycin. J. Am. Chem. Soc. **95**, 2352 (1973).
92. NAKAJIMA, M., M. INUKAI, T. HANEISHI, A. TERAHARA, M. ARAI, T. KINOSHITA, and C. TAMURA: Globomycin, a New Peptide Antibiotic with Spheroplast Forming Activity. III. Structural Determination of Globomycin. J. Antibiotics **31**, 426 (1978).
93. TSUKAGOSHI, N., G. TAMURA, and K. ARIMA: A Novel Protoplast-Bursting Factor (Surfactin) Obtained from *Bacillus subtilis* IAM 1213. I. The Effects of Surfactin on *Bacillus megaterium* KM. Biochim. Biophys. Acta **196**, 204 (1970).
94. ARIMA, K., KAKINUMA, A., and G. TAMURA: Surfactin, a Crystalline Peptidolipid Surfactant Produced by *Bacillus subtilis*: Isolation, Characterization and its Inhibition of Fibrin Clot Formation. Biochem. Biophys. Res. Comm. **31**, 488 (1968).
95. KAKINUMA, A., M. HORI, M. ISONO, G. TAMURA, and K. ARIMA: Determination of Amino Acid Sequence in Surfactin, a Crystalline Peptidolipid Surfactant Produced by *Bacillus subtilis*. Agric. Biol. Chem. **33**, 971 (1969).
96. KAKINUMA, A., M. HORI, H. SUGINO, I. YOSHIDA, M. ISONO, G. TAMURA, and K. ARIMA: Determination of the Location of Lactone ring in Surfactin. Agric. Biol. Chem. **33**, 1523 (1969).
97. KAKINUMA, A., A. OUCHIDA, T. SHIMA, H. SUGINO, M. ISONO, G. TAMURA, and K. ARIMA: Confirmation of the Structure of Surfactin by Mass Spectrometry. Agric. Biol. Chem. **33**, 1669 (1969).
98. KLUGE, B., J. VATER, J. SALNIKOV, and K. ECKART: Studies on the Biosynthesis of Surfactin, a Lipopeptide Antibiotic from *Bacillus subtilis* ATCC 21332. FEBS-Letters **231**, 107 (1988).
99. SHEPPARD, J.D., and C.N. MULLIGAN: The Production of Surfactin by *Bacillus subtilis* Grown on Peat Hydrolysate. Appl. Microbiol. Biotechnol. **27**, 110 (1987).
100. MINNIKIN, D.E., H. ABDOLRAHIMZADEH, and J. BADDILEY: Variation of Polar Lipid Composition of *Bacillus subtilis* (Marburg) with Different Growth Conditions. FEBS-Letters **27**, 16 (1972).
101. HOSONO, K., and H. SUZUKI: Acylpeptides, the Inhibitors of Cyclic Adenosine-3′,5′-monophosphate Phosphodiesterase. I. Purification, Physicochemical Properties and Structures of Fatty Acid Residues. J. Antibiotics **36**, 667 (1983).
102. HOSONO, K., and H. SUZUKI: Acylpeptides, the Inhibitors of Cyclic Adenosine-3′,5′-monophosphate Phosphodiesterase. II. Amino Acid Sequence and Location of Lactone Linkage. J. Antibiotics **36**, 674 (1983).
103. ITO, T. and H. OGAWA: Chemical Studies of the Antibiotic Esperin. The Structure of Esperin. Bull. Agric. Chem. Soc. Japan **23**, 536 (1959).
104. OVCHINNIKOV, Y.A., V.T. IVANOV, P.V. KOSTETSKY, and M.M. SHEMYAKIN: On the Structure of Esperin. L. Esperinic acid. Tetrahedron Letters 5285 (1966).
105. THOMAS, D.W., and T. ITO: The Revised Structure of the Peptide Antibiotic Esperin, Established by Mass Spectrometry. Tetrahedron **25**, 1985 (1969).

106. GUINAND, M., G. MICHEL, and E. LEDERER: Sur les lipides de *Nocardia asteroides*: isolement de lipopeptides. Compt. Rend. Acad. Sci. (Paris) **246**, 848 (1958).
107. GUINAND, M., and G. MICHEL: Structure d'un peptidolipide de *Nocardia asteroides*. Compt. Rend. Acad. Sci. (Paris) **256**, 1621 (1963).
108. GUINAND, M., G. MICHEL, and E. LEDERER: Structure de la peptidolipine NA. Compt. Rend. Acad. Sci. (Paris) **259**, 1267 (1964).
109. BARBER, M., W.A. WOLSTENHOLME, M. GUINAND, G. MICHEL, B.C. DAS, and E. LEDERER: Determination of Amino acid Sequences in Oligopeptides by Mass Spectrometry. II. The Structure of Peptidolipin NA. Tetrahedron Letters 1331 (1965).
110. MICHEL, G.: Lipopeptides des Mycobactéries et des Nocardia. In: Protides of the Biological Fluids, ed. by H. PEETERS, Elsevier Publ. Co., Amsterdam, 297 (1966).
111. PTAK, M., A. HEITZ, M. GUINAND, and G. MICHEL: A 400 MHz ^{1}H NMR Study of Peptidolipine NA, a Natural Cyclic Lipopeptide. Biochem. Biophys. Res. Comm. **94**, 1311 (1980).
112. GUINAND, M., M.J. VACHERON, G. MICHEL, B.C. DAS, and E. LEDERER: Détermination de séquences d'acides aminés dans des oligopeptides par la spectrométrie de masse. V. Structure de la Val^6-peptidolipine NA, nouveau peptidolipide de *Nocardia asteroides*. Tetrahedron, Suppl. 7, 271 (1966).
113. GUINAND, M., G. MICHEL, B.C. DAS, and E. LEDERER: Détermination de séquences d'acides aminés dans des oligopeptides par la spectrométrie de masse. VII. Structure de 1′ "α-amino-butyryl^1-peptidolipine NA", nouveau peptidolipide de *Nocardia asteroides*. Vietnamica Chim. Acta 37 (1966).
114. NISHII, M., T. KIHARA, and K. ISONO: The Structure of Lipopeptin A. Tetrahedron Letters **21**, 4627 (1980).
115. NISHII, M., K. ISONO, and K. IZAKI: Inhibition of Microbial Cell-Wall Synthesis by Lipopeptin A. Agric. Biol. Chem. **45**, 895 (1981).
116. SATOMI, T., H. KUSAKABE, G. NAKAMURA, T. NISHIO, M. URAMOTO, and K. ISONO: Neopeptins A and B, New Antifungal Antibiotics. Agric. Biol. Chem. **46**, 2621 (1982).
117. UBUKATA, M., M. URAMOTO, and K. ISONO: The Structures of Neopeptins, Inhibitors of Fungal Cell Wall Biosynthesis. Tetrahedron Letters **25**, 423 (1984).
118. OHNO, T., S. TAJIMA, and K. TOKI: Constitution of Viscosin. J. Agr. Chem. Soc. Japan **27**, 665 (1953).
119. HIRAMOTO, M., K. OKADA, S. NAGAI, and H. KAWAMOTO: Synthesis of the Proposed Structure of Viscosin. Biochem. Biophys. Res. Comm. **35**, 702 (1969).
120. HIRAMOTO, M., K. OKADA, and S. NAGAI: The Revised Structure of Viscosin, a Peptide Antibiotic. Tetrahedron Letters 1087 (1970).
121. BURKE, T.R., M. KNIGHT, and B. CHANDRASEKHAR: Solid-Phase Synthesis of Viscosin, a Cyclic Depsipeptide with Antibacterial and Antiviral Properties. Tetrahedron Letters **30**, 519 (1989).
122. SHOJI, J., and T. KATO: The Structure of Brevistin. J. Antibiotics **29**, 380 (1976).
123. BRECHT-FISCHER, A., H. ZÄHNER, and H. LAATSCH: Stoffwechselprodukte von Mikroorganismen. 183. Imacidin, ein neues Acylpeptidantibioticum. Arch. Microbiol. **122**, 219 (1979).
124. LAATSCH, H.: Metabolic Products of Microorganisms. 204. The Structure of Imacidin C. Liebigs Ann. Chem. 28 (1982).
125. DEBONO, M., M. BARNHART, C.B. CARRELL, J.A. HOFFMANN, J.L. OCCOLOWITZ, B.J. ABBOTT, D.S. FUKUDA, R.L. HAMIL, K. BIEMANN, and W.C. HERLIHY: A21978C, a Complex of New Acidic Peptide Antibiotics: Isolation, Chemistry and Mass Spectral Structure Elucidation. J. Antibiotics **40**, 761 (1987).
126. BOECK, L.D., D.S. FUKUDA, B.J. ABBOTT, and M. DEBONO: Deacylation of A21978C,

an Acidic Lipopeptide Antibiotic Complex, by *Actinoplanes utahensis*. J. Antibiotics **41**, 1085 (1988).

127. Debono, M., B.J. Abbott, R.M. Molloy, D.S. Fukuda, A.H. Hunt, V.M. Daupert, F.T. Counter, J.L. Ott, C.B. Carrell, L.C. Howard, L.D. Boeck, and R.H. Hamil: Enzymatic and Chemical Modifications of Lipopeptide Antibiotic A21978C; the Synthesis and Evaluation of Daptomycin. J. Antibiotics **41**, 1093 (1988).

128. Maget-Dana, R., J.H. Lakey, and M.Ptak: A Comparative Monomolecular Film Study of Antibiotic A21978C Homologues of Various Lipid Chain Length. Biochim. Biophys. Acta **962**, 201 (1988).

129. Thompson, R.Q., and M.S. Hughes: Stendomycin: a New Antifungal Antibiotic. J. Antibiotics, Ser. A, **16**, 187 (1963).

130. Bodanszky, M., J. Izdebski, I. Muramatsu, and A. Bodanszky: The Chemistry of the Peptide Antibiotic Stendomycin. Peptide 1968, North Holland Publ. Co., Amsterdam, 306 (1968).

131. Bodanszky, M., I. Muramatsu, and A. Bodanszky: Fatty Acid Constituents of the Antifungal Antibiotic Stendomycin. J. Antibiotics, Ser. A, **20**, 384 (1967).

132. Bodanszky, M., I. Muramatsu, A. Bodanszky, M. Lukin, and M.R. Doubler: Amino Acid Constituents of Stendomycin. J. Antibiotics **21**, 77 (1968).

133. Muramatsu, I., and M. Bodanszky: The Occurrence of Dehydro-butyrine in Stendomycin. J. Antibiotics **21**, 68 (1968).

134. Thomas, T.W., E. Lederer, M. Bodanszky, J. Izdebski, and I. Muramatsu: Partial Structure of the Peptide Antibiotic Stendomycin as Determined by mass Spectrometry. Nature **220**, 580 (1968).

135. Urry, D.W., and A. Ruiter: Conformation of Polypeptide Antibiotics .VI. Circular Dichroism of Stendomycin. Biochem. Biophys. Res. Commn., **38**, 800 (1970).

136. Peypoux, F., M. Guinand, G. Michel, L. Delcambe, B.C. Das, P. Varenne, and E. Lederer: Isolement de l'acide 3-amino-12-méthyl-tétradécanoïque et de l'acide 3-amino-12-méthyl-tridécanoïque à partir de l'iturine, antibiotique de *Bacillus subtilis*. Tetrahedron **29**, 3455 (1973).

137. Isogai, A., S. Takayama, S. Murakoshi, and A. Suzuki: Structures of β-amino Acids in Antibiotic Iturin A. Tetrahedron Letters **23**, 3065 (1982).

138. Peypoux, F., M. Guinand, G. Michel, L. Delcambe, B.C. Das, and E. Lederer: Structure of Iturin A, a Peptidolipid Antibiotic from *Bacillus subtilis*. Biochemistry **17**, 3992 (1978).

139. Nagai, U., F. Besson, and F. Peypoux: Absolute Configuration of an Iturinic Acid as Determined by CD Spectrum of its DNP-*p*-Methoxyanilide. Tetrahedron Letters 2359 (1979).

140. Garbay-Jaureguiberry, C., B.P. Roques, L. Delcambe, F. Peypoux, and G. Michel: NMR Conformational Study of Iturin A, an Antibiotic from *Bacillus subtilis*. FEBS-Letters **93**, 151 (1978).

141. Marion, D., M. Genest, A. Caille, F. Peypoux, G. Michel, and M. Ptak: Conformational Study of Bacterial Lipopeptides: Refinement of the Structure of iturin A in Solution by Two Dimensional ^{1}H-NMR and Energy Calculations. Biopolymers **25**, 153 (1986).

142. Winkelmann, G., H. Allgaier, R. Lupp, and G. Jung: Iturin A_L, a New Long Chain Iturin A Possessing an Unusual High Content of C_{16}-β-Amino acid. J. Antibiotics **36**, 1451 (1983).

143. Peypoux, F., F. Besson, G. Michel, L. Delcambe, and B.C. Das: Structure de l'iturine C de *Bacillus subtilis*. Tetrahedron **34**, 1147 (1978).

144. Besson, F., and G. Michel: Isolation and Characterization of New Iturins: Iturin D and Iturin E. J. Antibiotics **40**, 437 (1987).

145. LELIEVRE, D., and Y. TRUDELLE: Synthesis of the Cyclic Peptide of Iturin A. In: Second Forum on Peptides, ed. by A. AUBRY, M. MARRAUD and B. VITOUX, Colloque INSERM vol. **174**, 225 (1988).
146. PEYPOUX, F., M.T. POMMIER, B.C. DAS, F. BESSON, L. DELCAMBE, and G. MICHEL: Structures of Bacillomycin D and Bacillomycin L, Peptidolipid Antibiotics from *Bacillus subtilis*. J. Antibiotics **37**, 1600 (1984).
147. PEYPOUX, F., D. MARION, R. MAGET-DANA, M. PTAK, B.C. DAS, and G. MICHEL: Structure of Bacillomycin F, a New Peptidolipid Antibiotic of the Iturin group. Eur. J. Biochem. **153**, 335 (1985).
148. BESSON, F. and G. MICHEL: Bacillomycins F_b and F_c. Isolation and Characterization. J. Antibiotics **41**, 282 (1988).
149. PEYPOUX, F., G. MICHEL, and L. DELCAMBE: Structure de la mycosubtiline, antibiotique isolé de *Bacillus subtilis*. Eur. J. Biochem. **63**, 391 (1976).
150. PEYPOUX, F., M.T. POMMIER, D. MARION, M. PTAK, B.C. DAS, and G. MICHEL: Revised Structure of Mycosubtilin, a Peptidolipid Antibiotic from *Bacillus subtilis*. J. Antibiotics **39**, 636 (1986).
151. KATO, T., and J. SHOJI: The Amino Acid Sequence of Octapeptin C. J. Antibiotics **29**, 1339 (1976).
152. ROSENTHAL, K.S., P.E. SWANSON, and D.R. STORM: Disruption of *Escherichia coli* Outer Membranes by EM 49. A New Membrane Active Peptide Antibiotic. Biochemistry **15**, 5783 (1976).
153. DAS, B.C., and E. LEDERER: Mass spectrometry in Peptide Chemistry *In* New Techniques in Amino Acid, Peptide and Protein Analysis, ed. by A. NIEDERWIESER and G. PATAKI, Ann Arbor Science Publ. Inc., Ann Arbor, 175 (1971).
154. DAS, B.C., S.D. GERO, and E. LEDERER: N-Methylation of N-Acyl Oligopeptides. Biochem. Biophys. Res. Comm. **29**, 211 (1967).
155. DAS, B.C., S.D. GERO, and E. LEDERER: Detection and Localization of N-Methylaminoacid Residues in N-Acyl oligopeptides Methyl Esters by Mass Spectrometry. Nature **217**, 547 (1968).
156. THOMAS, D.W., B.C. DAS, S.D. GERO, and E. LEDERER: Advantages and Limitations of the Mass Spectrometric Sequence Determination of Permethylated Oligopeptide Derivatives. Biochem. Biophys. Res. Comm. **32**, 199 (1968).
157. Van HEIJENOORT, J., E. BRICAS, B.C. DAS, E. LEDERER, and W.A. WOLSTENHOLME: Détermination de séquences d'acides aminé dans des oligopeptides par la spectrométrie de masse. IX. Acylation avec de nouveaux radicaux mixtes; peptides contenant des acides aminés trifonctionnels. Tetrahedron **23**, 3403 (1967).
158. THOMAS, D.W.: Mass Spectrometry of Permethylated Peptide Derivatives: Extension of the Techniques to Peptides Containing Aspartic Acid, Glutamic Acid and Tryptophan. Biochem. Biophys. Res. Comm. **33**, 483 (1968).
159. BRICAS, E., J. Van HEIJENOORT, M. BARBER, W.A. WOLSTENHOLME, B.C. DAS, and E. LEDERER: Determination of Amino Acid Sequences in Oligopeptides by Mass Spectrometry. IV. Synthetic N-Acyl Oligopeptide Methyl Esters. Biochemistry **4**, 2254 (1965).
160. SCHULTEN, H.R.: Field Desorption Mass Spectrometry and Its Application to Biochemical Analysis, In: Methods of biochemical analysis, ed. by G. GLICK, John Wiley & Sons, New York, vol. **24**, 313 (1977).
161. HOWE, I., and M. JARMAN: New Techniques for the Mass Spectrometry of Natural Products, Progr. Chem. organ. natur. Products **47**, 107 (1985).
162. NAYLOR, S., and G. MONETI: Factors Affecting the Fragmentation of Peptides in Fast Atom Bombardment Mass Spectrometry. Biomed. Environm. Mass Spectrom. **18**, 405 (1989).

163. BESSON, F., F. PEYPOUX, G. MICHEL, and L. DELCAMBE: Mode of Action of Iturin A, an Antibiotic Isolated from *Bacillus subtilis*, on *Micrococcus luteus*. Biochem. Biophys. Res. Comm. **81**, 297 (1978).

164. BESSON, F., F. PEYPOUX, G. MICHEL, and L. DELCAMBE: Antifungal Activity upon *Saccharomyces cerevisiae* of Iturin A, Mycosubtilin, Bacillomycin L and of Their Derivatives. Inhibition of this antifungal activity by lipid antagonists. J. Antibiotics **32**, 828 (1979).

165. BESSON, F., F. PEYPOUX, M.J. QUENTIN, and G. MICHEL: Action of Antifungal Peptidolipids from *Bacillus subtilis* on the Cell Membrane of *Saccharomyces cervisiae*. J. Antibiotics **37**, 172 (1984).

166. QUENTIN, M.J., F. BESSON, F. PEYPOUX, and G. MICHEL: Action of Peptidolipidic Antibiotics of the Iturin Group on Erythrocytes. Effect of Some Lipids on Hemolysis. Biochim. Biophys. Acta **684**, 207 (1982).

167. QUENTIN, M. J., F. PEYPOUX, and G. MICHEL: Changes in Phospholipid Vesicles Size Induced by Amphipathic Antibiotic of the Iturin group. Biochem. intern. **7**, 63 (1983).

168. BESSON, F., F. PEYPOUX, and G. MICHEL: Action of Mycosubtilin and Bacillomycin L on *Micrococcus luteus* Cells and Protoplasts. Influence of the Polarity of the Antibiotics upon Their Action on the Bacterial Cytoplasmic Membrane. FEBS-Letters **90**, 36 (1978).

169. LATOUD, C., F. PEYPOUX, and G. MICHEL: Action of Iturin A, an Antifungal Antibiotic from *Bacillus subtilis*, on the Yeast *Saccharomyces cerevisiae*: Modifications of Membrane Permeability and Lipid Composition. J. Antibiotics **40**, 1588 (1987).

170. LATOUD, C., F. PEYPOUX, and G. MICHEL: Action of Iturin A on Membrane Vesicles from *Saccharomyces cervisiae*: Activation of Phospholipases A and B Activities by Picomolar Amounts of Iturin A. J. Antibiotics **41**, 1699 (1988).

171. MAGET-DANA, R., F. HEITZ, M. PTAK, F. PEYPOUX, and M. GUINAND: Bacterial Lipopeptides Induce Ion-Conducting Pores in Planar Bilayers. Biochem. Biophys. Res. Comm. **129**, 965 (1985).

172. MAGET-DANA, R., M. PTAK, F. PEYPOUX, and G. MICHEL: Effect of the O-Methylation of Tyrosine on the Pore-Forming Properties of Iturins. Biochim. Biophys. Acta **898**, 1 (1987).

173. PTAK, M., M. GENEST, D. MARION, R. MAGET-DANA, I. HARNOIS, D. GENEST, and A. CAILLE: Recent Progress in the Determination of Structure-Activity Relationships for a Family of Antifungal Lipopeptides In: Second Forum on Peptides, ed. by A. AUBRY, M. MARRAUD and B. VITOUS, Colloque INSERM n° 174, 11 (1988).

174. AYDIN, M., N. LUCHT, W. A. KÖNIG, R. LUPP, G. JUNG, and G. WINKELMANN: Structure Elucidation of the Peptide Antibiotics Herbicolin A and B. Liebigs Ann. Chem. 2285 (1985).

175. SMITH, D.W., and H.M. RANDALL: Mycosides of Mycobacteria. Amer. Rev. Respir. Diseases **92**, 34 (1965).

176. ASSELINEAU, C., and ASSELINEAU, J.: Waxes, Mycosides and Related Compounds. In: The Mycobacteria, a Source Book, ed. by G.P. KUBICA and L.G. WAYNE, Marcel Dekker Inc., New York, vol. 1, 345 (1984).

177. BRENNAN, P.J.: New Found Glycolipid Antigens of Mycobacteria. In: Microbiology-1984, ed. by L. LEIVE and D. SCHLESSINGER, Amer. Soc. Microbiol., Washington, 366 (1984).

178. JOLLES, P., F. BIGLER, T. GENDRE, and E. LEDERER: Sur la structure chimique du mycoside C, peptidoglycolipide de *Mycobacterium avium*. Bull. Soc. chim. biol. **43**, 177 (1961).

179. CHAPUT, M., G. MICHEL, and E. LEDERER: Structure du mycoside C_m, peptidoglycolipide de *Mycobacterium marianum*. Biochim. Biophys. Acta **63**, 310 (1962).

180. VILKAS, E., A. ROJAS, B.C. DAS, W.A. WOLSTENHOLME, and E. LEDERER: Détermination de séquences d'acides aminés dans des oligopeptides par la spectrométrie de masse. VI. Structure du mycoside C_b, peptidoglycolipide de *Mycobacterium butyricum*. Tetrahedron **22**, 2809 (1966).
181. LANEELLE, G., and J. ASSELINEAU: Structure d'un glycoside de peptidolipide isolé d'une mycobactérie. Eur. J. Biochem. **5**, 487 (1968).
182. DAFFE, M., M.A. LANEELLE, and G. PUZO: Structural Elucidation by Field Desorption and Electron-Impact Mass Spectrometry of the C-Mycosides Isolated from *Mycobacterium smegmatis*. Biochim. Biophys. Acta **751**, 439 (1983).
183. BRENNAN, P.J., and M.B. GOREN: Structural Studies on the Type-Specific Antigens and Lipids of the *Mycobacterium avium-M. intracellulare-M. scrofulaceum* serocomplex. *Mycobacterium intracellulare* serotype 9. J. Biol. Chem. **254**, 4205 (1979).
184. MACLENNAN, A.P.: The Monosaccharide Units in Specific Glycolipids of *Mycobacterium avium*. Biochem. J. **82**, 394 (1962).
185. BRUNETEAU, M., and G. MICHEL: Biogénèse des O-méthyl-6-désoxyhexoses présents dans le mycoside C_2. Biochim. Biophys. Acta **201**, 493 (1970).
186. VILKAS, E., A. ROJAS, and E. LEDERER: Sur un nouvel acide aminé, la N-méthyl O-méthyl L-serine, isolé des mycosides de *Mycobacterium butyricum* and *M. avium*. Compt. Rend. Acad. Sci. (Paris) **261**, 4258 (1965).
187. LANEELLE, G.: Sur la présence d'aminoalcools dans une fraction glyco-peptido-lipidique isolée d'une mycobactérie atypique. Compt. Rend. Acad. Sci. (Paris) Ser. C., **263**, 502 (1966).
188. LANEELLE, G.: Etude de deux formes lipophiles d'acides aminés produites par des mycobactéries. Thèse Doctorat-es-Sciences, Université de Toulouse, 1967.
189. VILKAS, E., E. LEDERER, and J.C. MASSOT: N-méthylation de peptides par la méthode d'Hakomori. Structure du mycoside C_{b1}. Tetrahedron Letters 3089 (1968).
190. VILKAS, E., C. GROS, and J.C. MASSOT: Sur la structure chimique d'un mycoside C isolé de *Mycobacterium scrofulaceum*. Compt. Rend. Acad. Sci. (Paris), Ser. C, **266**, 837 (1968).
191. VOILAND, A., M. BRUNETEAU, and G. MICHEL: Etude du mycoside C_2 de *Mycobacterium avium*. Détermination de la structure. Eur. J. Biochem. **21**, 285 (1971).
192. BRUNETEAU, M., and G. MICHEL: Synthèse du L-alaninol à partir d'extraits acellulaires de *Mycobacterium avium*. FEBS-Letters **14**, 57 (1971).
193. LANEELLE, G., J. ASSELINEAU, and G. CHAMOISEAU: Présence de mycosides C' (formes simplifiées de mycoside C) dans des bactéries isolées de bovins atteints de farcin. FEBS-Letters **19**, 109 (1971).
194. CAMPHAUSEN, R.T., R.L. JONES, and P.J. BRENNAN: A glycolipid Antigen Specific to *Mycobacterium paratuberculosis*. Structure and Antigenicity. Proc. Natl. Acad. Sci. USA, **82**, 3068 (1985).
195. TSANG, A.Y., V.L. BARR, J.K. MCCLATCHY, M. GOLDBERG, I. DRUPA, and P.J. BRENNAN: Antigenic Relationships of the *Mycobacterium fortuitum-Mycobacterium chelonae* complex. Intern. J. system. Bacteriol. **34**, 35 (1984).
196. DRAPER, P., and R.J.W. REES: The Nature of the Electron-Transparent Zone that Surrounds *Mycobacterium lepraemurium* Inside Host Cells. J. gen. Microbiol. **77**, 79 (1973).
197. BARROW, W.W., B.P. ULLOM, and P.J. BRENNAN: Peptidoglycolipid Nature of the Superficial Cell Wall Sheath of Smooth Colony-Forming Mycobacteria. J. Bacteriol. **144**, 814 (1980).
198. PORTAELS, F., M. DAFFE, M.A. LANEELLE, and C. ASSELINEAU: Etude de la composition lipidique de mycobactéries isolées de foies de tatous infectés par *Mycobacterium leprae*. Ann. Microbiol. (Inst. Pasteur) **135** A, 457 (1984).

199. BARROW, W.W., and P.J. BRENNAN: Isolation in High Frequency of Rough Variants of *Mycobacterium intracellulare* Lacking C-Mycoside Glycopeptidolipid Antigens. J. Bacteriol. **150**, 381 (1982).

200. SCHAEFER, W.B.: Serologic Identification and Classification of the Atypical Mycobacteria by their Agglutination. Amer. Rev. respir. Diseases **92** (suppl.), 85 (1965).

201. SCHAEFER, W.B.: Serologic Identification of the Atypical Mycobacteria and its Value in Epidemiologic Studies. Amer. Rev. respir. Diseases **96**, 115 (1967).

202. JENKINS, P.A.: Lipid Analysis in the Identification of Mycobacteria. An Appraisal. Rev. infect. Dis. **3**, 862 (1981).

203. BRENNAN, P.J., M. HEIFETS, and B.P. ULLOM: Thin-layer Chromatography of Lipid Antigens as a Means of Identifying Nontuberculous Mycobacteria. J. Clin. Microbiol. **15**, 447 (1982).

204. BRENNAN, P.J., H. MAYER, G.O. ASPINALL, and J.E. NAM SHIN: Structures of the Glycopeptidolipid Antigens from Serovars in the *Mycobacterium avium/M. intracellulare/M. scrofulaceum* Serocomplex. Eur. J. Biochem. **115**, 7 (1981).

205. CAMPHAUSEN, R.T., R.L. JONES, and P.J. BRENNAN: Structure and Relevance of the Oligosaccharide Hapten of *Mycobacterium avium Serotype* 2. J. Bacteriol. **168**, 660 (1986).

206. MCNEIL, M., A.Y. TSANG, and P.J. BRENNAN: Structure and Antigenicity of the Specific Oligosaccharide Hapten from the Glycopeptidolipid Antigen of *Mycobacterium avium* serotype A, the Dominant Mycobacterium Isolated from Patients with Acquired Immune Deficiency Syndrome. J. Biol. Chem. **262**, 2630 (1987).

207. BRENNAN, P.J., G.O. ASPINALL, and J.E. NAM SHIN: Structures of the Specific Oligosaccharides from the Glycopeptidolipid Antigens of Serovars in the *Mycobacterium avium/M. intracellulare/M. scrofulaceum* Complex. J. Biol. Chem. **256**, 6817 (1981).

208. BOZIC, C.M., M. MCNEIL, D. CHATTERJEE, I. JARDINE, and J.P. BRENNAN: Further Novel Amidosugars Within the Glycopeptidolipid Antigens of *Mycobacterium avium*. J. Biol. Chem. **263**, 14984 (1988).

209. MCNEIL, M., H. GAYLORD, and P.J. BRENNAN: N-Formyl Kansosaminyl-(1 → 3)-2-O-methyl-D-rhamnopyranose: The Type-Specific Determinant of Serovariant 14 of the *Mycobacterium avium* Complex. Carbohyd. Res. **177**, 185 (1988).

210. CHATTERJEE, D., C. BOZIC, G.O. ASPINALL, and P.J. BRENNAN: Glucuronic Acid- and Branched Sugar-Containing Glycolipid antigens of *Mycobacterium avium*. J. Biol. Chem. **263**, 4092 (1988).

211. CHATTERJEE, D., G.O. ASPINALL, and P.J. BRENNAN: The presence of Novel Glucuronic Acid-Containing, Type-Specific Glycolipid Antigens Within *Mycobacterium* spp. Revision of Earlier Structures. J. Biol. Chem. **262**, 3528 (1987).

212. GERWIG, G.J., J.P.KAMERLING, and J.F.G. VLIEGENHART: Determination of the D and L Configurations of Neutral Monosaccharides by High-Resolution Capillary G.L.C. Carbohyd. Res. **62**, 349 (1978).

213. HUNTER, S.W., T. FUJIWARA, R.C. MURPHY, and P.J. BRENNAN: N-Acyl Kansosamine. A Novel Acylamino Sugar from the Trehalose-Containing Lipooligosaccharide Antigens of *Mycobacterium Kansasii*. J. Biol. Chem. **259**, 9729 (1984).

214. TERELITSKY, M.J., and W.W. BARROW: Postphagocytic Detection of Glycopeptidolipids Associated with the Superficial L_1 Layer of *Mycobacterium intracellulare*. Infect. & Immun. **41**, 1312 (1983).

215. BRENNAN, P.J.: Antigenic Peptidoglycolipids, Phospholipids and Glycolipids, *In* The Mycobacteria, a Source Book, ed. by G.P. KUBICA and L.G. WAYNE, Marcel Dekker Inc., New York, 467 (1984).

216. WOODBURY, J.L., and W.W. BARROW: Radiolabelling of *Mycobacterium avium* Oligosaccharide Determinant and Use in Macrophage Studies. J. Gen. Microbiol. **135**, 1875 (1989).
217. DAVID, H.L., N. RASTOGI, S. CLAVEL-SERES, and F. CLEMENT: Alterations in the Outer Wall Architecture Caused by the Inhibition of Mycoside C Biosynthesis in *Mycobacterium avium*. Curr. Microbiol. **17**, 61 (1988).
218. RASTOGI, N., V. LEVY-FREBAULT, M.C. BLOM-POTAR, and H.L. DAVID: Ability of Smooth and Rough Variants of *Mycobacterium avium* and *M. intracellulare* to Grow and Survive Intracellularly. Role of C-Mycosides. Zbl. Bakt. Hyg. A, **270**, 345 (1989).
219. FURUCHI, A., and T. TOKUNAGA: Nature of the Receptor Substance of *Mycobacterium smegmatis* for D4 Bacteriophage Adsorption. J. Bacteriol. **111**, 404 (1972).
220. GOREN, M.B., J.K. MCCLATCHY, B. MARTENS, and O. BROKL: Mycosides C: Behavior as Receptor Site Substance for Mycobacteriophage D4. J. Virol. **9**, 999 (1972).
221. DHARIWAL, K.R., A. LIAV, A.E. VATTER, G. DHARIWAL, and M.B. GOREN: Haptenic Oligosaccharides in Antigenic Variants of Mycobacterial C-Mycosides Antagonize Lipid Receptor Activity for Mycobacteriophage D4 by Masking a Methylated Rhamnose. J. Bacteriol. **168**, 283 (1986).
222. MOORE, R.E., V. BORNEMANN, W.P. NIEMCZURA, J.M. GREGSON, J.L. CHEN, T.R. NORTON, G.M.L. PATTERSON, and G.L. HELMS: Puwainaphycin C, a Cardioactive Cyclic Peptide from the Blue-Green Algae *Anabaena* BQ 16-1. Use of Two Dimensional ^{13}C-^{13}C and ^{13}C-^{15}N Correlation Spectroscopy in Sequencing the Amino Acid Units. J. Am. Chem. Soc. **111**, 6128 (1989).
223. SEGRE, A., R.C. BACHMANN, A. BALLIO, F. BOSSA, I. GRGURINA, N.S. IACOBELLIS, G. MARINO, P. PUCCI, M. SIMMACO, and J.Y. TAKEMOTO: The Structure of Syringomycins A_1, E and G. FEBS-Letters **255**, 27 (1989).

(*Received October 3, 1989*)

Naturally Occurring Di- and Trithiophenes

J. KAGAN, Department of Chemistry, University of Illinois at Chicago, Chicago, Illinois, U.S.A.

With 1 Figure

Contents

I. Introduction

The dithiophenes and trithiophenes reviewed in this article are part of a large group of biogenetically related molecules found in plants of the family Compositae (Asteraceae). They include compounds having a variable number of unsaturations, particularly double bonds and triple bonds, which occur singly or in combinations. The first report of a naturally occurring trithiophene, α-terthienyl in the flowers of *Tagetes erecta*, appeared in 1947 (*270*); the first naturally occurring dithiophene was isolated from *Bidens radiata* and described in 1961 (*142*). The book *Naturally Occurring Acetylenes*, by BOHLMANN *et al.* (*29*), is a superb review of the field up to 1972. Interestingly, out of its more than 500 pages, only two were devoted to physiological and pharmacological aspects. BOHLMANN and ZDERO later contributed one chapter, "Naturally Occurring Thiophenes", to a volume *Thiophenes and its Derivatives* which appeared in 1985 (*57*). This chapter presents a survey of such thiophenes based on biogenetic considerations, includes an extensive analysis of the distribution of these compounds, and discusses methods of analysis based on UV, ^{1}H-NMR, ^{13}C-NMR, and mass spectra. The most recent references in the chapter came from publications appearing in 1981.

In view of the information on naturally occurring dithiophenes and trithiophenes already available from these sources, this survey emphasizes the occurrence, the biosynthesis, the synthesis, the photophysical and photochemical properties, and especially the biological properties of these molecules.

II. Nomenclature

In all naturally occurring polythiophenes found to date the thiophene rings are attached through the 2- and 5-positions, next to the sulfur atoms, which are also known as α-positions. In most naturally occurring dithiophenes the two thiophene rings are directly connected but in a few cases the rings are separated by either two or four carbon atoms. While the older name for a molecule having two attached thiophene rings was bithienyl, or more specifically α-bithienyl, the current terminology favors 2,2′-bithiophene. In contrast, in all the trithiophenes known the three rings are connected without intervening atoms. When three thiophene rings are connected through their α-positions, the earlier nomenclature favored names based on terthienyl, or more specifically α-terthienyl.

However, since this leads to awkward naming of substituted derivatives, the preferred terminology for the parent compound is now 2,2′:5′,2″-trithiophene. In this manner, now used in *Chemical Abstracts*, each position in each ring is easily identified through the prime and double prime symbols. Mercifully, the largest α-thiophene oligomer synthesized and definitely characterized to date contains only seven thiophene rings.

III. Naturally Occurring Dithiophenes: Structures

Structures of naturally occurring dithiophenes with their Chemical Abstracts registry numbers are listed arbitrarily in Tables 1–7 according to their substitution patterns.

IV. Distribution of Dithiophenes

The distribution of naturally occurring dithiophenes is detailed in Table 8, in which the genera are listed in alphabetical order. All genera

Table 1. *Dithiophenes of Type* H–(2-thienyl)–(2-thienyl)–C≡C—R

R		No.	CAS No.
R = CH=CH$_2$		**1**	1134-61-8
C(Cl)=C(H)—OAc		**2**	14892-16-1
C(Cl)=CH(OAc)		**3**	14744-57-1
COCH$_2$OAc		**4**	1222-83-9
C≡C—CH$_3$		**5**	61102-17-8
CH(R$_1$)—CH$_2$(R$_2$)			
R$_1$ = H	R$_2$ = OH	**6**	1137-87-7
H	OAc	**7**	1219-28-9
H	OiVal	**8**	61102-15-6
OH	OH	**9**	1211-45-6
OH	OAc	**10**	95910-62-6
OH	iVal	**11**	58930-56-6
OH	Cl	**12**	1020-03-7
OAc	OH	**13**	1687-88-3
OAc	OAc	**14**	1233-95-0
OAc	Cl	**15**	96850-15-6

Table 2. *Dithiophenes of Type* R_1–(2-thienyl)–(2-thienyl)–$C{\equiv}C{-}R_2$

R_1	R_2	No.	CAS
$R_1 = CH_3$	$R_2 = CH{=}CH_2$	**16**	1137-83-3
CH_2OH	$CH{=}CH_2$	**17**	1211-41-2
CH_2OAc	$CH{=}CH_2$	**18**	1152-21-2
CH_2OiBu	$CH{=}CH_2$	**19**	27123-31-5
$CH_2OAngel$	$CH{=}CH_2$	**20**	26901-18-8
CH_2OTigl	$CH{=}CH_2$	**21**	26944-52-5
CH_2OSen	$CH{=}CH_2$	**22**	26905-74-8
CHO	$CH{=}CH_2$	**23**	61102-16-7
CH_3	CH_2CH_2OH	**24**	58930-55-5
CH_3	CH_2CH_2OAc	**25**	58930-54-4
CH_3	$CH(OAc){-}CH_2OAc$	**26**	58930-57-7
CH_3	$CH(OH){-}CH_2Cl$	**27**	78876-54-7
CH_2OH	CH_3	**28**	102054-38-6
CHO	CH_3	**29**	102054-37-5
COOH	CH_3	**30**	32155-99-0
$C{\equiv}CH$	CH_3	**31**	17257-07-7
$CH{=}CH_2$	CH_3	**32**	17257-06-6
$COCH_3$	CH_3	**33**	102054-40-0
$COCH_2OH$	CH_3	**34**	102054-41-1
$COCH_2OAc$	CH_3	**35**	
COCOOH	CH_3	**36**	102054-36-4
$COCOOCH_3$	CH_3	**37**	102054-42-2
$CH(OH){-}COOH$	CH_3	**38**	102054-35-3
$CH(OH){-}CH_2OH$	CH_3	**39**	119875-32-0
CH_2OAc	CH_2CH_3	**40**	89913-52-0
CH_2OAc	CH_2CH_2OAc	**41**	77363-67-8
CH_2OiVal	CH_2CH_2OiVal	**42**	96853-77-9
CH_2OTigl	CH_2CH_2iVal	**43**	
CH_2OSen	CH_2CH_2OiVal	**44**	114849-21-7

$(CH_3)_2CHCH_2COOH$ iso-Valeric

$H(CH_3)C{=}C(CH_3)COOH$ **Tiglic**

$H(CH_3)C{=}C(CH_3)COOH$ **Angelic**

$(CH_3)_2C{=}CHCOOH$ **Senecic**

Table 3. *Dithiophenes of Type* R_1–(thiophene)–(thiophene)–R_2

R_1 = H	R_2 = $COCH_3$	**45**	3515-18-2
H	$CH{=}CH{-}CH_2CH_2OAc$	**46**	92202-49-8
CH_3	$CH{=}CH{-}CH{=}CH_2(E)$	**47**	17257-08-8
$CH_2{=}CH$	$CH{=}CH{-}CH_3(E)$	**48**	1518-19-0

Note: the compound proposed to be **46** was subsequently proved to be **7** instead (*36*, *12*, *40*).

Table 4. *Dithiophenes of Type* (thiophene)–C≡C–(thiophene)–C≡C–R

R = CH_3	**49**	80750-43-2
CH_2OH	**50**	36687-71-5
CH_2OAc	**51**	36687-70-4
CHO	**52**	36687-72-6

Table 5. *Dithiophenes of Type* (thiophene)–C≡C–(thiophene)–R

R = CHO	**53**	36687-75-9
CH_2OH	**54**	36687-74-8
CH_2OAc	**55**	36687-73-7
$COCH_3$	**56**	36687-90-8
$CHOHCH_3$	**57**	36687-89-5
$CH{-}CH_3$ (OAc)	**58**	36687-91-9
$CHOHCH_2OH$	**59**	

containing di- and trithiophenes are found in the family Compositae, even though acetylenic compounds have been found in several other plant families and in algae, marine organisms and microorganisms. No attempts have been made to group plant genera according to tribes, as these alliances may not have permanent status. The list combines the results of a computer search of Chemical Abstracts for the literature since 1967 with the references found in (*29*) and (*57*). To our knowledge, some of the then unpublished data (*30*, *31*) have remained unpublished. The gender and spelling of some names has changed, and at least one species appears to have been cited under more than one binomial. No attempt has been made to evaluate critically the taxonomic determinations.

Table 6. *Dithiophenes of Type*

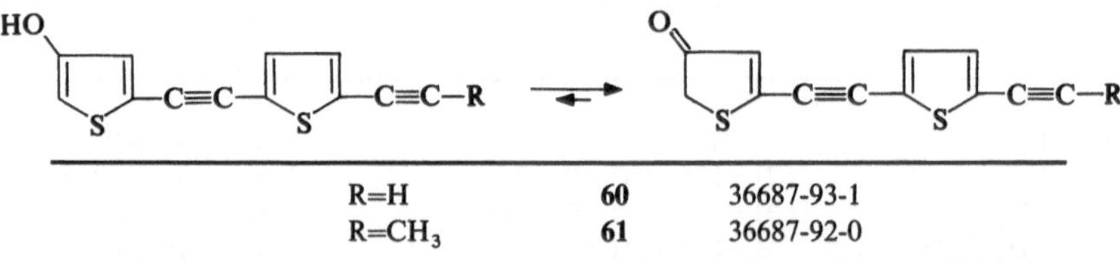

R=H	**60**	36687-93-1
R=CH_3	**61**	36687-92-0

Table 7. *Dithiophenes of Type*

E	**62**	112051-66-8
Z	**63**	112051-65-7
	62 + **63**	55249-45-1

Table 8. *Distribution of Naturally Occurring Dithiophenes*

Adenophyllum porophylloides	**1, 6, 10**	(*99*)
Aphyllocladus denticulatus	**29**	(*176*)
Arnica sachalinensis	**1**	(*31*)
Arctium lappa	**28, 29, 30, 33, 34, 35, 36, 37, 38, 39**	(*193, 252*)
Bellida graminae	**1**	(*31*)
Berkheya		
B. adlamii (= *B. radula*)	**1, 2**	(*59*)
B. angustifolia	**1**	(*47*)
B. armata	**49, 50, 51, 53, 54, 56, 57, 58**	(*52*)
B. barbata	**14, 60, 61**	(*52*)
B. bergiana	**1, 2, 3, 14**	(*45*)
B. bipinnatifida	**1, 6, 49, 50**	(*41, 45*)
B. carduoides	**62, 63**	(*47*)
B. carlinopsis	**1, 4, 6, 7, 14,**	(*44*)
B. cirsifolia	**1**	(*45*)
B. coriacea	**1, 6, 7, 14**	(*47*)
B. debilis	**1, 6, 14**	(*45*)
B. decurrens	**49, 51, 52**	(*52*)
B. echinata	**1, 2, 3, 49**	(*45*)
B. fruticosa	**10, 13, 14**	(*52*)
B. erysithales	**1, 7, 14, 49, 50**	(*45*)
B. herbacea	**49, 50, 51, 52, 53, 54, 55**	(*52*)
B. heterophylla	**1, 6, 10, 14, 49, 50**	(*52*)
B. insignus	**1, 6, 7, 14**	(*45*)
B. macrocephala	**1, 2**	(*52*)
B. maritima	**2, 3, 14**	(*45*)
B. multijuga	**1, 6, 7, 14**	(*45*)

Table 8. (Contd)

B. onopordifolia	**1, 2**	(*52*)
B. pannosa	**6, 7**	(*45*)
B. pauciflora	**6, 7, 14**	(*44*)
B. purpurea	**49, 51, 53**	(*52*)
B. radula (= *B. adlamii*)	**1, 2, 6, 9**	(*41, 52*)
B. rhapontica	**1, 2, 6, 7, 9, 14**	(*41, 45*)
B. rigida	**49, 50, 53, 56**	(*31*)
B. robusta	**1, 3**	(*45*)
B. setifera	**49, 50, 51**	(*45*)
B. speciosa	**1, 2, 14**	(*45*)
B. umbellata	**49, 50,**	(*45*)
B. zeyheri	**18, 40,**	(*43*)
B. sp. novum aff. bipinnatifida	**1, 7, 14, 49, 50**	(*45*)
Bidens		
B. connata	**1, 48**	(*22*, 25)
B. dahlioides	**17, 18**	(*25*)
B. frondosa	**48**	(*30*)
B. radiata	**47**	(*142*)
Buphthalmum		
B. grandifolium	**16, 17, 18**	(*27*)
B. salicifolium	**16, 17, 18**	(*27, 51*)
Calea		
C. angusta	**17, 18**	(*35*)
C. pilosa	**16, 18, 38**	(*32*)
Cardopatium corymbosum	**1** and dimers, **6, 7, 14**	(*31, 217*)
Centaurea		
C. carduiformis	**1**	(*31*)
C. depressa	**1**	(*31*)
C. sphaerocephala	**59**	(*17*)
Chrysactinia mexicana	**1, 6, 7, 10, 14**	(*95, 99*)
Cullumia		
C. bisulca	**1, 49**	(*52*)
C. decurrens	**1, 7, 14**	(*31*)
C. setosa	**1, 6, 10, 14**	(*52*)
C. squarrossa	**1, 5, 6, 7, 9, 14, 49**	(*47*)
C. sulcata	**1, 6, 14, 49**	(*47*)
Cuspida cernua	**1, 6, 10, 14**	(*52*)
Didelta		
D. carnosa	**1, 6, 10, 14, 49**	(*52*)
D. spinosa	**10, 14**	(*52*)
Dyssodia		
D. acerosa	**5, 6, 7, 15, 23**	(*60*)
D. anthemidifolia	**1, 6, 7, 10, 17, 18**	(*53, 99, 102*)
D. decipiens	**1, 6, 7, 16**	(*56, 102*)
D. papposa	**1, 6, 7**	(*60, 102*)
D. setifolia	**1, 6, 7, 9, 11, 14, 18, 24, 25, 26**	(*53*)
Echinops		
E. bannaticus	**1, 6, 7, 9, 10, 13, 14, 17, 18**	(*24*)
E. champtavicus	**1, 6, 7, 9, 10, 13, 14, 17, 18**	(*30*)

Table 8. (Contd)

E. commutatus	**1, 6, 7, 9, 10, 13, 14, 17, 18**	(*24*)
E. cornigerus	**1, 6, 7, 9, 10, 13, 14, 17, 18**	(*24*)
E. dahuricus	**1, 6, 7, 9, 10, 13, 14, 17, 18**	(*24*)
E. exaltus	**1, 6, 7, 9, 10, 13, 14, 17, 18**	(*30*)
E. horridus	**1, 6, 7, 9, 10, 13, 14, 17, 18**	(*24*)
E. humilis	**1, 6, 7, 9, 10, 13, 14, 17, 18**	(*30*)
E. niveus	**1, 6, 7, 9, 10, 13, 14, 17, 18**	(*24*)
E. persicus	**1, 6, 7, 9, 10, 13, 14, 17, 18**	(*24*)
E. ritro	**1, 6, 7, 9, 10, 13, 14, 17, 18**	(*24*)
E. sphaerocephalus	**1, 6, 7, 9, 10, 13, 14, 17, 18**	(*24*)
E. spinosissimus	**1**	(*31*)
E. strigosus	**1, 6, 7, 9, 10, 13, 14, 17, 18**	(*24*)
E. viscosus	**1, 6, 7, 9, 10, 13, 14, 17, 18**	(*30*)
Eclipta		
E. alba	**1, 12, 17, 18, 19, 20, 21, 22**	(*50*)
E. erecta	**1, 12, 16, 17, 18, 19, 20, 21, 22, 42, 43, 44**	(*50, 136, 220, 221*)
E. prostrata (published as *E. prostata*)	**1, 12, 17, 18, 19, 20, 21**	(*50*)
Flaveria		
F. australasica	**1, 17, 18**	(*31*)
F. bidentis	**1, 9**	(*31*)
F. brownii	**1, 6**	(*100*)
F. campestris	**1, 14, 18**	(*31*)
F. chloraefolia	**1, 9, 14, 18**	(*42*)
F. linearis	**1**	(*100*)
F. pringlei	**1, 6, 7, 14, 18**	(*31, 100*)
F. repanda	**18**	(*39*)
F. trinervata	**18**	(*31*)
F. trinervis	**1, 6, 7, 18**	(*30, 100*)
Guizotia oleifera	**33**	(*34*)
Haploesthes greggii	**1, 6, 7, 8, 45**	(*60*)
Helichrysum trilineatum	**14**	(*58*)
Hymenatherum		
H. acerosa	**1, 6, 10**	(*99*)
H. pentachaeta	**1, 6, 10**	(*99*)
H. tenuifolium	**1, 6, 7**	(*56*)
Mutisia		
M. coccinea	**1, 14**	(*54*)
M. homoeantha	**1, 7**	(*62*)
Nicolletia trifida	**1, 10**	(*99*)
Platycarpha glomerata	**49, 50, 51**	(*55*)
Porophyllum		
P. gracile	**1, 6, 7, 10**	(*99, 102*)
P. lanceolatum	**1, 6, 7**	(*56, 74*)
P. riedelii	**1**	(*61*)
P. ruderale	**1, 14, 25, 41**	(*38, 56, 99*)
P. scoparia	**1, 6, 11, 14, 15**	(*26*)
P. scoparium	**1, 6, 10**	(*99*)

Table 8. (Contd)

Pterocaulon virgatum	**12**, **28**	(*23*)
Ratibida columnifera	**16**	(*122*)
Rudbeckia amplexicaulis	**16**, **17**, **18**, **47**	(*34*, *40*)
Schoenia cassiniana	**1**, **18**	(*31*)
Tagetes		
T. coronopifolia	**1**, **6**, **7**	(*31*, *74*, *102*)
T. decipiens	**1**	(*56*)
T. elliptica	**1**	(*31*, *74*)
T. erecta	**1**, **4**, **6**, **7**, **32**, **33**, **46**	(*34*, *36*, *82*, *102*, *117*, *123*)
T. filifolia	**1**, **7**	(*31*, *74*, *102*)
T. glandulifera = *T. minuta*		(*12*)
T. gracilis	**1**, **6**, **7**	(*56*)
T. indica	**1**, **6**, **7**	(*31*)
T. lemmonii	**1**, **6**, **7**	(*31*, *74*, *102*)
T. lucida	**1**, **7**	(*30*, *74*)
T. microglossa	**1**, **6**, **7**	(*70*, *71*, *188*)
T. minuta	**1**, **6**, **7**, **12**, **16**	(*11*, *12*, *31*, *74*, *82*, *102*, *117*, *124*)
T. multiflora	**1**	(*102*)
T. patula	**1**, **4**, **6**, **7**, **9**, **10**, **14**, **32**, **33**, **46**	(*36*, *56*, *76*, *102*, *105*, *121*, *123*, *142*, *144*, *192*, *197*, *229*, *233*)
T. signata	**1**	(*31*, *74*)
T. tenuifolia	**1**, **6**	(*102*)
T. terniflora	**1**, **6**, **7**	(*56*)
T. zypaquirensis	**1**	(*56*)
Tessaria integrifolia	**1**	(*64*, *74*, *137*)
Tymophylla tenuiloba	**1**, **6**, **7**	(*56*)
Viguiera stenoloba	**5**	(*63*)
Wedelia forsteriana	**16**, **17**, **18**	(*51*)

V. Naturally Occurring Trithiophenes: Structures

Structures of naturally occurring trithiophenes are listed in Table 9. Their distribution is listed in Table 10.

VI. Distribution of Trithiophenes

Only one trithiophene, α-terthienyl (**T1**), has been found in a large number of species. The known distribution of the others is very restricted. With the exception of **T2**, their occurrence seems to be restricted to single

Table 9. *Trithiophenes of Type* [structure: terthiophene with R_1 on the 3′ position and R_2 on the 5″ position]

T1	$R_1 = H$	$R_2 = H$	1081-34-1
T2	$R_1 = H$	$R_2 = CH_3$	26905-73-7
T3	$R_1 = H$	$R_2 = CH_2OH$	13059-93-3
T4	$R_1 = H$	$R_2 = CH_3OCOCH_3$	26905-77-1
T5	$R_1 = H$	$R_2 = CH_2OCO{-}C(CH_3){=}C(H){-}CH_3$	27123-32-6
T6	$R_1 = H$	$R_2 = CH_2OCO{-}C(CH_3){=}C(CH_3){-}H$	26901-19-9
T7	$R_1 = H$	$R_2 = CH_2OCOCH(CH_3)_2$	26905-75-9
T8	$R_1 = H$	$R_2 = CH_2OCOCH{=}C(CH_3)_2$	26905-76-0
T9	$R_1 = H$	$R_2 = CHO$	7342-41-8
T10	$R_1 = OCH_3$	$R_2 = H$	58930-53-3

Table 10. *Distribution of Naturally Occurring Trithiophenes*

Adenophyllum porophylloides	**T1, T2**	(*99*)
Berkheya		
B. adlamii See *B. radula*		
B. angustifolia	**T1**	(*47*)
B. barbata	**T1**	(*52*)
B. bipinnatifida	**T1**	(*45*)
B. carlinopsis	**T1**	(*44*)
B. carduoides	**T1**	(*47*)
B. cirsifolia	**T1**	(*45*)
B. echinata	**T1**	(*45*)
B. fructicosa	**T1**	(*52*)
B. herbacea	**T1**	(*52*)
B. heterophylla	**T1**	(*52*)
B. macrocephala	**T1**	(*52*)
B. maritima	**T1**	(*45*)
B. onopordifolia	**T1**	(*52*)
B. pannosa	**T1**	(*45*)
B. radula (= *B. adlamii*)	**T1**	(*52, 59*)
B. rhapontica	**T1**	(*41, 45*)
B. robusta	**T1**	(*45*)
B. setifera	**T1**	(*45*)
B. speciosa	**T1**	(*45*)
B. umbellata	**T1**	(*45*)
Cardopatium corymbosum	**T1**	(*31, 217, 224*)

Table 10. (Contd)

Centaurea		
C. depressa	**T1**	(*31*)
C. kotschyi	**T1**	(*31*)
Chrysactinia mexicana	**T1**	(*95*)
Cullumia		
C. setosa	**T1**	(*52*)
C. sulcata	**T1**	(*47*)
C. squarrosa	**T1**	(*47*)
Cuspidia cernua	**T1**	(*52*)
Didelta		
D. carnosa	**T1**	(*52*)
D. spinosa	**T1**	(*52*)
Dyssodia		
D. acerosa	**T1, T2, T9**	(*60*)
D. anthemidifolia	**T1, T2, T10**	(*53, 99, 102*)
D. decipiens	**T1, T10**	(*56, 102*)
D. papposa	**T1, T10**	(*60, 102*)
Echinops		
E. bannaticus	**T1**	(*24*)
E. chamtavicus	**T1**	(*31*)
E. commutatus	**T1**	(*24*)
E. cornigerus	**T1**	(*24*)
E. dahuricus	**T1**	(*23*)
E. exaltus	**T1**	(*30*)
E. humilis	**T1**	(*30*)
E. horridus	**T1**	(*24*)
E. niveus	**T1**	(*24*)
E. persicus	**T1**	(*24*)
E. ritro	**T1**	(*24*)
E. sphaerocephalus	**T1**	(*24*)
E. spinosissimus	**T1**	(*31*)
E. strigosus	**T1**	(*24*)
E. viscosus	**T1**	(*30*)
Eclipta		
E. alba	**T1, T2, T3, T4, T5, T6, T7, T8, T9**	(*50, 168, 169*)
E. erecta	**T1, T2, T3, T4, T5, T6, T7, T8, T9**	(*50, 136, 220*)
E. prostrata (published as *E. prostata*)	**T1, T2, T3, T4, T5, T6, T7, T8, T9**	(*50*)
Erato polymnoides (= *Munnozia polymnoides*)	**T1**	(*33*)
Flaveria		
F. bidentis	**T1**	(*31*)
F. brownii	**T1, T2**	(*100*)
F. campestris	**T1**	(*31*)
F. chloraefolia	**T1**	(*42*)
F. linearis	**T1, T2**	(*100*)
F. pringlei	**T1**	(*31, 100*)
F. repanda	**T1**	(*39, 46*)
F. trinervata	**T1**	(*46*)
F. trinervis	**T1, T2**	(*100*)

Table 10. (Contd)

Gaillarida pulchella	**T1**	(*30*)
Haploestes greggii	**T1**	(*60*)
Helichrysum		
H. splendidum	**T1**	(*48*)
H. polycladum	**T1**	(*58*)
Hymenatherum		
H. acerosa	**T1, T2**	(*99*)
H. pentachaeta	**T1, T2**	(*99*)
Munnozia polymnoides		
(= *Erato polymnoides*)	**T1**	(*33*)
Mutisia homoeantha	**T1**	(*62*)
Nicolletia trifida	**T1**	(*99*)
Porophyllum		
P. gracile	**T1**	(*99, 102*)
P. lanceolatum	**T1**	(*56, 74*)
P. riedelii	**T1**	(*61*)
P. ruderale	**T1**	(*56, 99*)
P. scoparia	**T1**	(*26*)
Tagetes		
T. coronopifolia	**T1**	(*31, 74, 102*)
T. elliptica	**T1**	(*31, 74*)
T. erecta	**T1**	(*34, 36, 82, 102, 117, 123, 270*)
T. filifolia	**T1**	(*31, 74*)
T. gracilis	**T1**	(*56*)
T. indica	**T1**	(*30*)
T. jalisciencis	**T1**	(*188*)
T. lemmonii	**T1**	(*31, 74, 102*)
T. lucida	**T1**	(*30*)
T. microglossa	**T1**	(*70, 71, 188*)
T. minuta	**T1**	(*31, 74, 82, 102, 117, 124*)
T. multiflora	**T1**	(*102*)
T. patula	**T1, T2**	(*36, 56, 76, 102, 121, 123, 142, 144, 192, 229, 230, 233*)
T. signata	**T1**	(*30*)
T. tenuifolia	**T1**	(*31, 74, 102*)
Tessaria integrifolia	**T1**	(*74*)
Thymophylla tenuiloba	**T1**	(*56*)

genera. The other trithiophenes might be present in many more plants, but their low concentration has allowed them to escape analysis. Alternatively, their biosynthesis might be under much stricter genetic control, making them valuable taxonomic markers.

VII. Biosynthesis of Di- and Trithiophenes

Surprisingly little work has been done on the biosynthesis of bi- and trithiophenes. The steps involved in the formation of such plant components start with the synthesis of the carbon skeleton, introduce sulfur atoms which are then incorporated into thiophene rings, and finish with minor side-chain adjustments. These may include reduction, oxidation, formation of epoxides, conversion of the latter into diols, esterification of hydroxyl groups, loss of carbon atoms, etc.

As in all biosynthetic investigations, determining the origin of the key carbon, hydrogen, and sulfur atoms in a molecule is a major undertaking. Just as important, the enzyme systems involved in each step need to be isolated and characterized and their specificity determined. There is now some understanding of the biogenesis of thiophene derivatives, but little knowledge of the detailed steps in the biosynthesis of most of the members of the family. However, recently introduced techniques for growing cultures of root tissues in well defined media promise to help elucidate many of the biosynthetic steps for these natural products and to work out their enzymology.

An interesting question is whether the observed synthesis of a given compound is "natural", or whether it occurs in response to variable external challenges. For example, the synthesis of polyacetylenic compounds can be induced in cell cultures of *Bidens pilosa* by fungal culture filtrates (*93*). The same effect has also been observed in *B. sulphureus* and in unspecified *Tagetes* cultures (*105*). The extent to which some of the presumed normal thiophene constituents of plants may actually be the result of a phytoalexin response is not known.

VII. 1. Experimental Difficulties

The formulation of any biosynthetic pathway depends on the isolation or at least on the detection of all the presumed intermediates. Polyacetylenic compounds are often unstable molecules whose successful isolation depends on the techniques used in handling both the plants and their extracts. The possible influence of work-up procedures on the yields of isolated components has been recently investigated with thin-layer chromatography over silica gel, a standard analytical and preparative procedure in natural products analysis. The recovery of **1** and α-terthienyl (**T1**) placed on silica gel plates ranged between 77% and 97%

immediately after their application (*171*). When the extraction was performed 24 h after application, the recovery ranged from 43 to 74% for **T1** but only from 18 to 57% for **1**. This study suggests strongly that more labile compounds are likely to be even more readily affected by "mild" silica gel chromatography and that this procedure may yield unreliable quantitative (and perhaps even qualitative) data. Interestingly, admixture of several substances with the silica gel significantly protected the thiophenes from degradation. Although the degradation of plant components was attributed to light-dependent reactions, experiments did not clearly support this explanation, for example by comparing the degradation of samples kept in light with controls kept in the dark.

Another recent study investigated how the handling of seedlings of *Tagetes patula* affected the detection of α-terthienyl, bithiophenes, and benzofuran derivatives (*230*). The thiophene components were **1**, **6**, **7**, **9**, **14**, and **T1**. Seedlings of the same age were (a) deep-frozen (−80°C), (b) freeze-dried, (c) dried at 40°C in the dark, (d) dried at 40°C in the presence of incandescent light, or (e) dried at 105°C in the dark. Analysis of the extracts by hplc showed that procedure (a) led to the highest amounts of extractable components. As expected, drying at high temperature was the most deleterious, the benzofuran and the minor component **9** all but disappearing after this treatment. Surprisingly, the amount of **6** detected per mg of dry weight was the lowest in procedure (a), the values being in the ratios of 1, 1.6, 2.7, 2.4, and 1.9 respectively in going from (a) to (e). Clearly the alcohol must be, at least partially, an artefact of the isolation procedures. Compound **9** was also more abundant in the freeze-dried sample than in the freshly frozen sample, but disappeared on heating.

In the same study, exposure of the crude plant extracts to daylight and to near-ultraviolet light resulted in a time-dependent change in the concentration of the components. As expected from LAM and THOMASEN's study (*171*), irradiation produced a dramatic change in the hplc peak intensities. Less than half of **1** remained after 3 h exposure to daylight, and the disappearance was almost total after 40 min of exposure to UV. Again, the concentration of **6** was greater after daylight exposure and did not greatly diminish upon UV treatment.

In most studies, the effect of the isolation procedure and the handling of the extracts on the composition of the mixture of secondary metabolites are assessed with less care. One should therefore be somewhat skeptical about postulates of biological or evolutionary relationships which are based solely on the presence or absence of specific compounds in published analyses of plants.

VII. 2. Biosynthesis of the Carbon Backbone

In 1953, CHALLENGER and HOLMES wrote that "it may be more than a coincidence that the only instance so far recorded of the occurrence of a true thiophene derivative in plants should be in a family so many members of which contain polyacetylenes. The α-terthienyl may arise by interaction of H_2S with a straight chain compound containing an acetylenic-olefinic system. . . . Other reactions such as oxidation, decarboxylation, or dehydrogenation might be involved. It could be argued that a long-chain paraffin or fatty acid might serve equally well as the starting point." (*72*) The picture of the biosynthesis of thiophenes which has since developed generally agrees with these predictions.

BOHLMANN and his collaborators made major contributions to postulates regarding the biosynthesis of naturally occurring acetylenic compounds and their derivatives, and have expressed their views in their highly significant book, *Naturally Occurring Acetylenes* (*29*). However, a critical reading of the various publications in this area leads to the conclusion that surprisingly little concrete biosynthetic evidence has been acquired experimentally and that extensive extrapolations have provided current notions on the biosynthesis of bi- and trithiophenes.

Feeding experiments have determined that polyacetylenic compounds resulted from the incorporation of malonyl-CoA and acetyl-CoA; oleic acid was established as a key precursor. A number of carbon atoms must be discarded in going from a precursor with 18 carbon atoms to some of the simpler dithiophenes, many with 12 and one with just 10

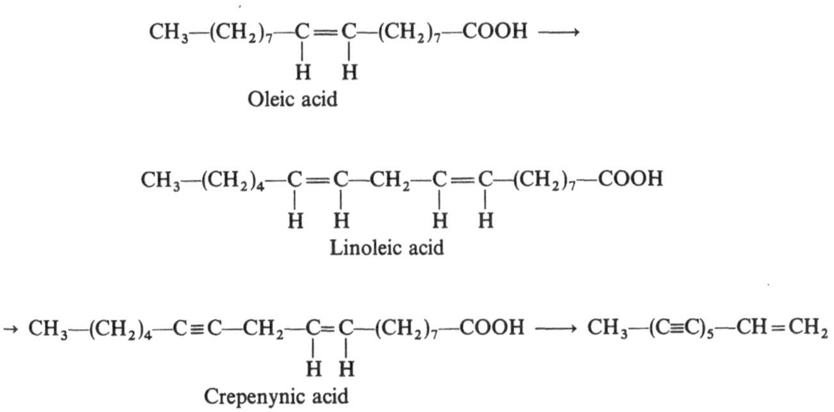

$$CH_3{-}(CH_2)_7{-}CH{=}CH{-}(CH_2)_7{-}COOH \longrightarrow$$

Oleic acid

$$CH_3{-}(CH_2)_4{-}CH{=}CH{-}CH_2{-}CH{=}CH{-}(CH_2)_7{-}COOH$$

Linoleic acid

$$\rightarrow CH_3{-}(CH_2)_4{-}C{\equiv}C{-}CH_2{-}CH{=}CH{-}(CH_2)_7{-}COOH \longrightarrow CH_3{-}(C{\equiv}C)_5{-}CH{=}CH_2$$

Crepenynic acid

Scheme 1. The conversion of oleic acid into a polyacetylene carbon atoms. Whether the losses occur before or after the formation of the thiophene rings is not known.

Further oxidation of oleic to linoleic acid (which contains one additional non-conjugated double bond) and then to crepenynic acid (which introduces the first triple bond) eventually leads to the tridecapentaynene shown in Scheme 1, thought to be a more immediate precursor to all the dithiophenes and trithiophenes discussed here. Since bi- and trithiophenes are often found in plants with polyacetylenic compounds, a close biological relationship between these compounds is certainly plausible.

VII. 3. Labelling Experiments

In a little known work, PALASZEK examined the conversion of several simple ^{14}C-labelled precursors into α-terthienyl in *T. erecta* (*196*). These were DL-glucose-UL-^{14}C, DL-ornithine-2-^{14}C, sodium malonate-2-^{14}C, DL-cystine-1-^{14}C, DL-serine-3-^{14}C, malonic-2-^{14}C acid, DL-methionine-2-^{14}C, and sodium pimelate-7-^{14}C. The dilution factors for the last three were the lowest, 1865, 1385 and 445 respectively. Thus preformed polyacetylenes are not necessarily required.

All the feeding experiments with less distant precursors reported in the literature are listed in Scheme 2 and Table 11.

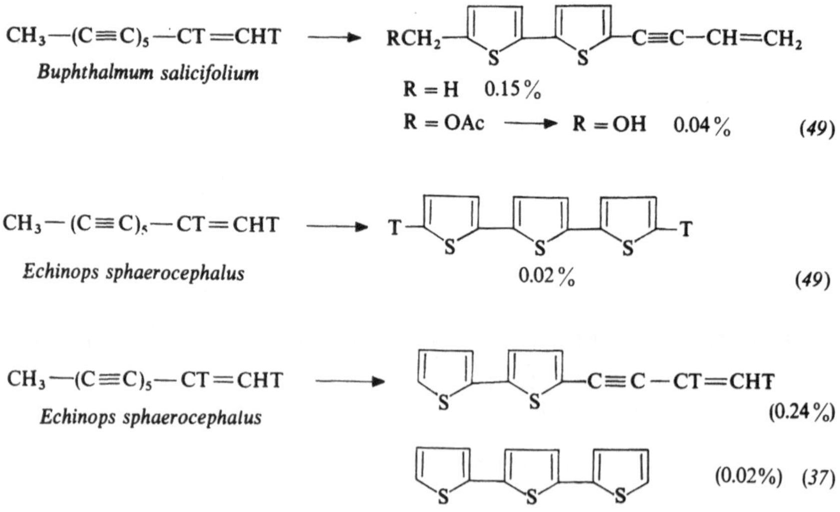

$CH_3—(C{\equiv}C)_5—CT{=}CHT$ ⟶ RCH_2–[thiophene]–[thiophene]–$C{\equiv}C—CH{=}CH_2$

Buphthalmum salicifolium

R = H 0.15%

R = OAc ⟶ R = OH 0.04% (*49*)

$CH_3—(C{\equiv}C)_5—CT{=}CHT$ ⟶ T–[thiophene]–[thiophene]–[thiophene]–T

Echinops sphaerocephalus 0.02% (*49*)

$CH_3—(C{\equiv}C)_5—CT{=}CHT$ ⟶ [thiophene]–[thiophene]–$C{\equiv}C—CT{=}CHT$ (0.24%)

Echinops sphaerocephalus

[thiophene]–[thiophene]–[thiophene] (0.02%) (*37*)

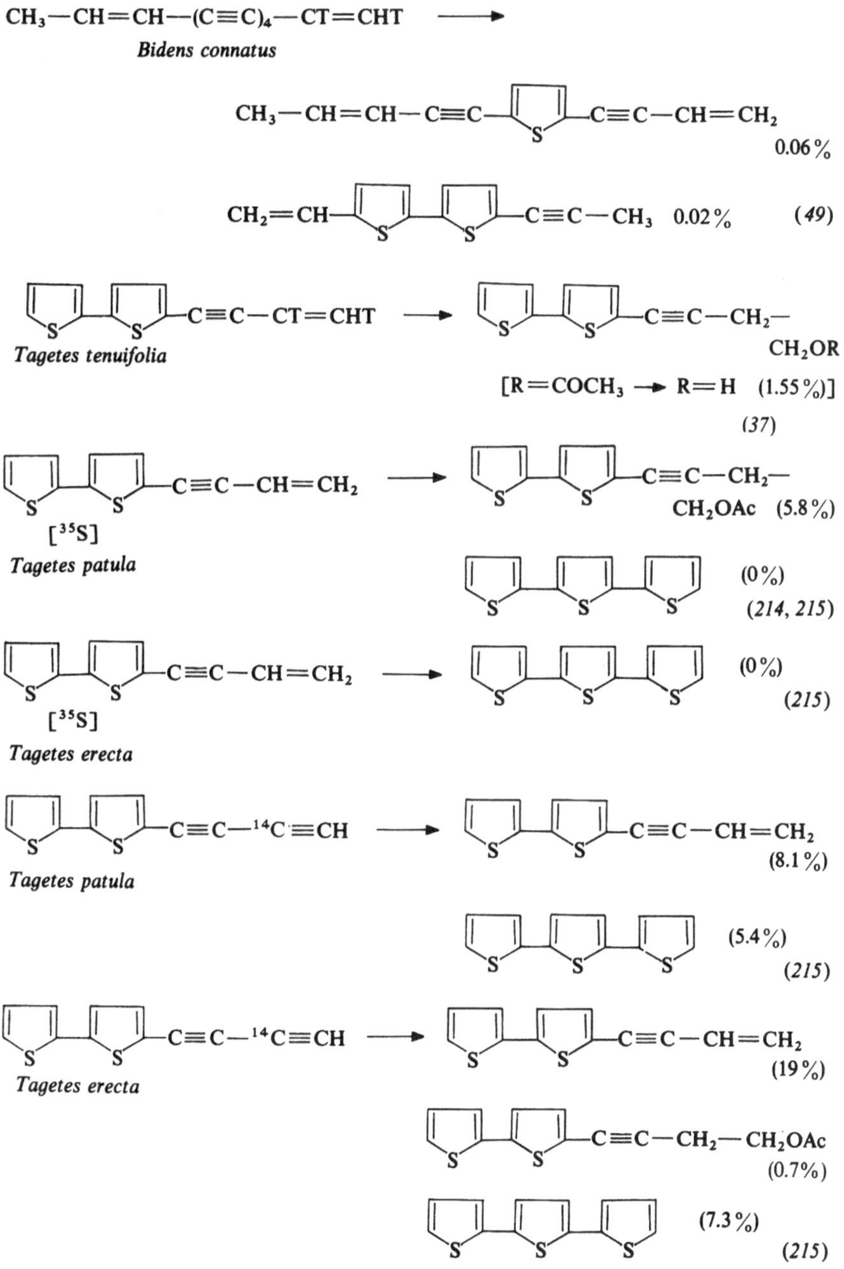

$CH_3—CH{=}CH—(C{\equiv}C)_4—CT{=}CHT$ ⟶

Bidens connatus

$CH_3—CH{=}CH—C{\equiv}C$— S —$C{\equiv}C—CH{=}CH_2$ 0.06%

$CH_2{=}CH$— S — S —$C{\equiv}C—CH_3$ 0.02% (*49*)

S — S —$C{\equiv}C—CT{=}CHT$ ⟶ S — S —$C{\equiv}C—CH_2—CH_2OR$

Tagetes tenuifolia

[$R{=}COCH_3$ → $R{=}H$ (1.55%)]

(*37*)

S — S —$C{\equiv}C—CH{=}CH_2$ ⟶ S — S —$C{\equiv}C—CH_2—CH_2OAc$ (5.8%)

[^{35}S]

Tagetes patula

S — S — S (0%)

(*214*, *215*)

S — S —$C{\equiv}C—CH{=}CH_2$ ⟶ S — S — S (0%)

[^{35}S]

Tagetes erecta

(*215*)

S — S —$C{\equiv}C—^{14}C{\equiv}CH$ ⟶ S — S —$C{\equiv}C—CH{=}CH_2$ (8.1%)

Tagetes patula

S — S — S (5.4%)

(*215*)

S — S —$C{\equiv}C—^{14}C{\equiv}CH$ ⟶ S — S —$C{\equiv}C—CH{=}CH_2$ (19%)

Tagetes erecta

S — S —$C{\equiv}C—CH_2—CH_2OAc$ (0.7%)

S — S — S (7.3%)

(*215*)

Scheme 2. Feeding experiments with radioactive precursors leading to thiophene derivatives

JENTE *et al.* investigated how a precursor with 13 carbon atoms was converted into a product with 12 carbon atoms (*144*). They assessed the conversion of two series of bithiophenes, one with 12 and the other with 13 carbon atoms, into products in which the side chain had been modified, and into α-terthienyl. The compounds tested are shown in Scheme 3. Each of these compounds was tested in *Tagetes patula* for incorporation into the four products shown below (Scheme 4). The percent incorporations into each product tested are summarized in Table 11:

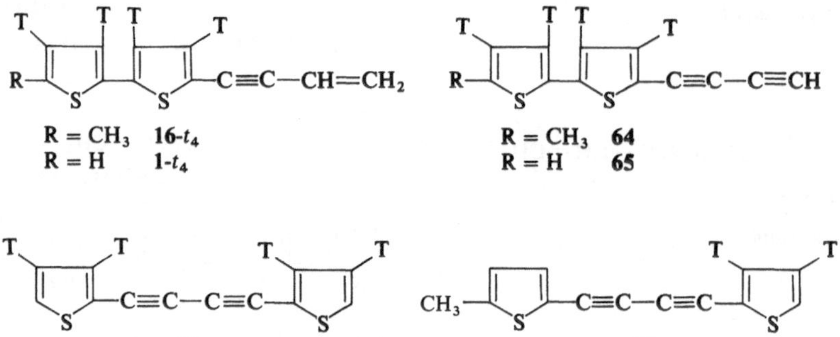

$R = CH_3$ **16-t_4**
$R = H$ **1-t_4**

$R = CH_3$ **64**
$R = H$ **65**

66

67

Scheme 3. Tritiated precursors fed to *Tagetes patula*

6 $R = H$
7 $R = Ac$

T1 $R = H$
T2 $R = CH_3$

Scheme 4. Components of *Tagetes patula* tested for tritium incorporation

Table 11. *Results of the feeding experiments in Tagetes patula*

	T1	**6 and 7**	**T2**
from **16-t_4**	0.001	0.132	≥ 0.006
from **1-t_4**	0.005	0.470	
from **64**	1.250	0.002	≥ 0.012
from **65**	1.160	0.002	
from **66**	0.385	0	
from **67**	0.066	0	≥ 0.011

Finally, the most recent feeding experiments with *Tagetes patula* involved a mixture of the compounds **1**-t_4 and **65**. The plants were assayed for **6** and **7** and **T1**, and the incorporation values remained nearly constant over a period of several months (*142*).

All the experiments mentioned above were performed on just six different plants. It is not yet known whether the biosynthetic pathways are absolutely identical in all the plants. If so, the following conclusions may be reached:

a) The polyunsaturated compounds tested, with four or five triple bonds and one or two double bonds, are precursors for bi- and trithiophenes, but the efficiency of the incorporations is always below 0.25%. The question is whether such low values reflect slow transport of precursors into the plants, allowing competitive synthesis and extensive dilution in the pool of unlabelled products, or whether the values are low because they do not result from direct biosynthetic pathways.

b) The most efficient reactions uncovered were the conversions of dithienylbutadiyne into the partially reduced enyne product **6** (8.1% efficiency) and into α-terthienyl in *T. patula* and *T. erecta* (5.4% and 7.3% respectively). These fairly comparable numbers suggest that both compounds are formed in parallel rather than sequential paths. This conclusion is supported by an experiment in *Tagetes patula* which used the ^{35}S-labelled enyne and yielded inactive α-terthienyl. Since uptake and incorporation of the labelled compound in this experiment were proven by the activity in the acetate, the negative result is significant. It is also supported by the data of Jente *et al.* (*144*).

The conclusion that a dithienylbutadiyne is a good precursor for α-terthienyl is also supported by feeding experiments in *Tagetes patula* (*144*), which further indicate that the presence or absence of an additional methyl group does not effect the conversion into α-terthienyl. This result suggests that perhaps the same enzyme complex performs the overall conversion of the methyl derivative. However, there is no information on whether the thirteenth carbon atom is lost before or after the last thiophene ring is formed. The fact that α-terthienyls substituted with methyl and hydroxymethyl groups have been isolated suggests that, at least in the plants studied, the loss of the extra carbon atom can occur after completion of the trithiophene system. It is not known whether 12- and 13-carbon precursors are handled differently in different plants. One must note that the butadiyne symmetrically substituted with two thiophene rings is also converted into α-terthienyl in *T. patula*, as shown in the case of **66** (Table 1), and that in this case the conversion is definitely more facile when an extra methyl group (as in **67**) is absent.

The experiments performed with ^{3}H- and ^{35}S-labelled molecules did not give the same results when **1** was used as precursor to α-terthienyl. The former case yielded the low level of incorporation of 0.005% (Table 11), whereas no incorporation was found in the latter case (Scheme 2). The question of whether the tritium incorporation reflects incomplete purification of the compound, reincorporation of tritiated fragments into the biosynthetic pool, or real but inefficient biosynthesis has not been addressed experimentally. No ^{14}C-labelled precursor has been used in this specific conversion.

No feeding experiments have elucidated the sequence in which the thiophene rings are formed from acyclic precursors. The products isolated already had at least two thiophene rings. The only exception is in the work in *Bidens connatus*, where tritiated products with one and two thiophene rings were isolated. However, the possible intermediacy of the monothiophene in the biosynthesis of the bithiophene was not investigated.

Analytical limitations probably explain the paucity of information concerning the changes in constituents and/or in their relative concentrations during the course of the plants' development. The advent of hplc techniques removed this limitation, at least partially. As illustrated by a study of the distribution of thiophene derivatives in different organs in *T. patula* seedlings grown under different conditions (*228*), the technique demonstrated high turnover rates for each of the compounds during the course of the development.

VII. 4. Incorporation of Sulfur

The detailed path(s) leading to incorporation of sulfur into thiophenes are not known. Successful incorporation of ^{35}S into thiophene rings of both inorganic and organic precursors has been recorded. These were Na_2SO_4 (*196*, *213*), methionine (*213*), and cysteine (*196*, *215*). Both approaches were preparatively useful for obtaining sulfur-35 labelled products used in feeding experiments, but Na_2SO_4 was incorporated more efficiently than methionine when administered through the roots. In contrast NaHS was not incorporated (*213*). Such experiments may not provide a true picture of the biosynthesis, because they combine at least two variables, namely the uptake and transport of the precursor to the site(s) of biosynthesis, and the incorporation of a usable form of sulfur into the final thiophenes. The low incorporation of sodium sulfide may owe as much to its toxicity to the plants (*96*) as to the biosynthesis itself.

PALASZEK felt that cysteine was clearly the precursor of the sulfur atom in **1** and **T1** (*196*).

Circumstantial evidence suggests that butadiynes may be precursors to thiophenes. The model reaction in which a 1,3-butadiyne is treated with Na_2S to yield a thiophene, originally suggested by CHALLENGER and HOLMES (*72*), can be easily carried out *in vitro*. However, there are not yet any persuasive data on the reaction *in vivo*. The lack of incorporation of NaHS does not rule out the formation of a sulfide by an enzyme complex involved with the biosynthesis of a thiophene ring. It has been argued that two separate steps may be involved in the addition, unlike the easy laboratory cyclization process which directly produces a thiophene derivative. The first step would produce an enyne species potentially capable of cyclizing to a thiophene. Such an intermediate may have been trapped as its methyl thioether, as shown in Scheme 5 for *Flaveria repanda* (*37*).

$$CH_3{-}(C{\equiv}C)_5{-}CT{=}CHT \longrightarrow CH_3{-}(C{\equiv}C)_2{-}\underset{\displaystyle SCH_3}{\underset{|}{C}}{=}CH{-}(C{\equiv}C)_2{-}CH{=}CH_2$$

Flaveria repanda (0.01%)

Scheme 5. Sulfur introduction into an acyclic polyacetylene in *Flaveria repanda*

However, a 1,4-dithione might also be formed upon addition of a second equivalent of H_2S onto one adjacent triple bond. In the laboratory, the reaction of a 1,4-butanedione with H_2S is a standard method of synthesis of thiophenes. It has not been tested as a pathway for the conversion of diacetylenes into thiophenes or as a more direct alternative to 1,3-butadiynes in the biosynthesis of thiophenes.

VII. 5. 1,2-Dithiins

Oxidative cyclization of the enolic form of the above 1,4-butanedithione could lead to the antiaromatic six-membered ring product possessing two adjacent sulfur atoms (a 1,2-dithiin). Such structures, named thiarubrenes because of their red color, have been found only in Compositae, and only in the genera *Ambrosia*, *Chaenactis*, *Eriophyllum*, and *Rudbeckia* (*80*, *191*). When heated or irradiated, thiarubrenes lose sulfur and produce thiophenes. The mechanism of this formal extrusion reaction has not been elucidated. One possibility involves the intermediacy of the isomeric enedithione. Alternatively, the loss of sulfur might occur through a 1,1-dioxide formed by oxidation, followed by loss of SO_2. It is not known whether 1,2-dithiins are intermediates in the biosynthesis of thiophenes in those genera where they have been found, and their intermediacy in the biosynthesis of thiophenes in the other genera has not been specifically eliminated.

BOHLMANN has suggested that H_2S_2, rather than H_2S, might be the sulfur-introducing reagent (*28*). In this postulate (Scheme 6) a double

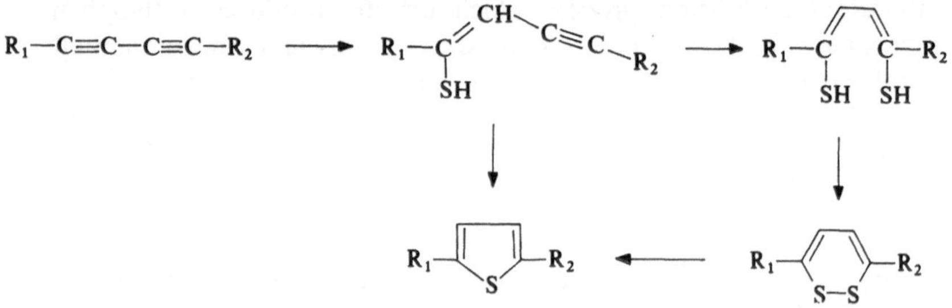

R_1—C≡C—C≡C—R_2 → R_1—C(SH)=CH—C≡C—R_2 → R_1—C(SH)=CH—CH=C(SH)—R_2

Scheme 6. Formal routes from 1,3-butidyines to thiophenes and 1,2-dithiins

addition reaction of a thiol moiety onto an acetylene would give the thiarubrene, which would then lose one sulfur atom to produce the final thiophene ring. An alternative to the cyclization which forms a 6-membered ring (a 6-endo-dig process) could be cyclization to form a 5-membered ring isomer (a 5-exo-dig process). Neither process is disfavored in Baldwin's classification (*15*). Interestingly, when the reaction of a 1,3-butadiyne with Na_2S_2 was tested in the laboratory, it produced a thiophene in addition to the 5-membered ring disulfide from the 5-exo-dig reaction (Scheme 7). Although the authors concluded that the

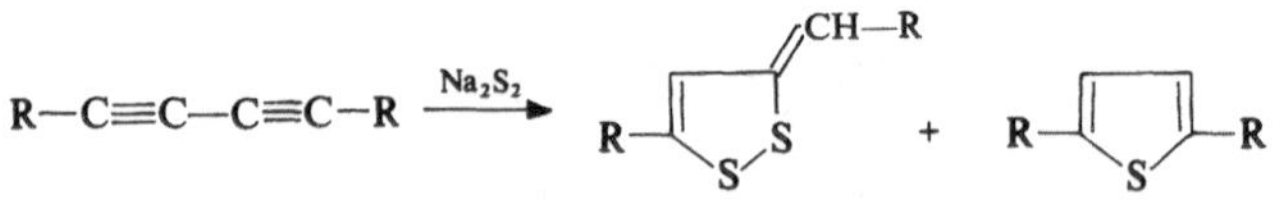

R—C≡C—C≡C—R $\xrightarrow{Na_2S_2}$ (5-membered ring disulfide with =CH—R) + (2,5-disubstituted thiophene)

Scheme 7. The reaction of Na_2S_2 with a 1,3-butadiyne

disulfide addition reaction was not a likely mode of synthesis for thiarubrenes *in vivo* (*28*) this conclusion requires additional experimental support since the thiophene isolated from the reaction *in vitro* may actually indicate that the desired thiarubrene had been formed, but had decomposed under the reaction conditions.

VII. 6. Enzymatic Studies

Isolation and characterization of the enzymes involved provides the ultimate touch to the formulation of biosynthetic pathways. Here again, the field is still in its infancy. Almost nothing is known about the enzymology of the pathway(s) leading to di- and trithiophenes, or about their regulation. Enzymes involved in side chain modifications of bithiophenes with a four carbon side chain were the first to be investigated. A highly specific 5-(4-acetoxy-1-butynyl)-2,2′-bithiophene: acetate esterase occurring in the aerial parts of *T. patula* was partially purified (*227*). It catalyzes the reaction shown in Scheme 8, but it could not be decided whether this enzyme was involved in a degradative or a biosynthetic step.

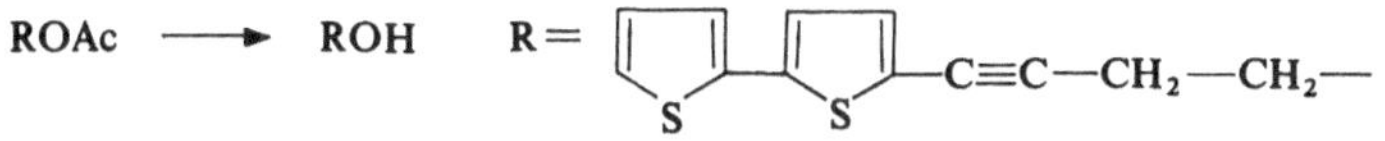

ROAc ⟶ ROH R= (2,2′-bithienyl)–C≡C—CH_2—CH_2—

Scheme 8

Further careful analysis of seedlings of *T. patula* revealed the presence of a diacetate congener. An enzyme system different from the esterase of Scheme 8 which hydrolyzes the diacetate to a diol was recognized and partially purified (Scheme 9). The kinetic data implicated cooperation of two enzymes. Interestingly, the activity of this enzyme system was found to change over time in hypocotyls, with the highest activity in the

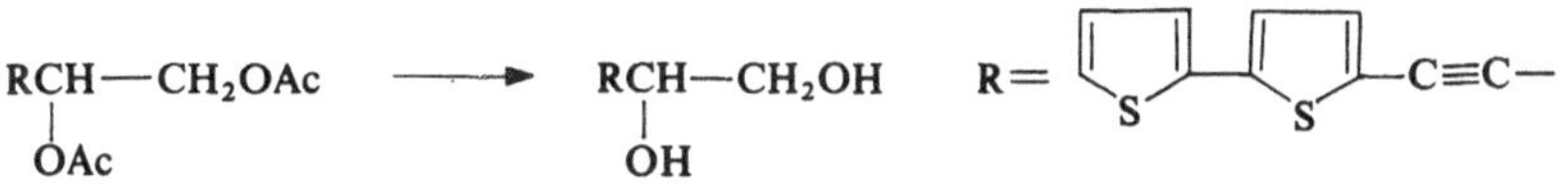

RCH(OAc)—CH_2OAc ⟶ RCH(OH)—CH_2OH R= (2,2′-bithienyl)–C≡C—

Scheme 9

youngest plants. In the roots, the enzyme activity remained constant (*200*). Later studies in *T. patula* seedlings also identified acetyl-CoA : hydroxybutynyl-bithiophene O-acetyltransferases, enzymes involved in the reverse steps of acetylating the monoalcohol or the diol (*186*). These enzymatic studies were reviewed by Sütfeld (*229*) who suggested that some of the thiophene constituents of *Tagetes patula*, such as the alcohols and their esters, actively participate in the biochemical development of the plants, while others are end products. His observation that **1** and **T1** occur almost exclusively in senescing tissues like fading cotyledons, flower heads or in ripening achenes led him to suggest that these two components were end products.

Similar studies were reported by TOSI *et al.* (*233*) who used hplc, tlc and hptlc techniques to determine the distribution of thiophene derivatives at 13 stages between germination and senescence. Flowering plants contained the highest total amount of thiophenes. While the concentration of the other constituents varied with age, that of the monoalcohol **6** and of **T1** were constant. They concluded that bithiophene synthesis takes place mainly in the roots and that trithiophene synthesis takes place mainly in the leaves.

No enzymes associated with the early stages of the biosynthesis of di- and trithiophenes have so far been purified or characterized.

VII. 7. Genetic Analyses

Only one report has been published dealing with a genetic analysis in the synthesis of thiophene derivatives. Since plants of the genus *Flaveria* have been found to assimilate CO_2 by C_3 and C_3–C_4 intermediate pathways in addition to the C_4 mechanism, DOWNUM *et al.* wondered whether links could be found between these photosynthetic mechanisms and the patterns of thiophene production (*100*). The following plant materials were analyzed: *F. pringlei* (C_3), *F. linearis* (C_3–C_4), *F. brownii* and *F. trinervis* (C_4), as well as the hybrids *F. pringlei* × *F. brownii* ($C_3 \times C_4$), and *F. trinervis* × *F. linearis* and *F. brownii* × *F. linearis* ($C_4 \times C_3$–C_4). Minor differences were observed, perhaps caused by differences in the age of the plants and other environmental variables. In any event, the hybrids did produce the thiophenes **1**, **6**, **7**, **T1**, and **T2**, which had no decipherable relationship to the thiophenes in the parents. There appear to have been no other genetic studies using di- or trithiophenes as markers.

VII. 8. Tissue Cultures

In principle, use of tissue cultures in well defined media under controlled conditions (illumination, temperature, etc.) represents a technique ideal for obtaining reproducible information on plant biochemistry, undisturbed by climatic and ecological variables. One important question is whether undifferentiated cells produce the same enzyme systems and the same secondary metabolites as intact plants. The answer probably depends on the compound(s) of interest and on the culture conditions. Recent work illustrates the importance of this research area, whose successes promise to provide new tools for the study of biosynthetic pathways leading to polythiophenes.

Tissue cultures frequently do not produce the compounds of interest present in the plant or they no longer do so after several serial transfers. As described by Croes *et al.* for *T. erecta* and *T. minuta* (*82*), callus cells from the uppermost internodes of the stem can be grown in standard media. Their production of dithiophenes and α-terthienyl strongly depended on the level of illumination and could be varied widely by supplementing the medium with growth hormones. The callus cultures of the two species differed markedly. There were no signs of morphological differentiation in *T. minuta*, but roots regenerated in the cell cultures of *T. erecta* after about 3 weeks, and this coincided with a sharp increase in the levels of dithiophenes and α-terthienyl. Some root and shoot formation also occurred in *T. patula* callus cultures (*166*).

Calli of *T. patula* subcultured one or two times contained **1** and **7** but the concentrations were only about one-third in the second subculture. In contrast, liquid cultures started from these released thiophenes, mostly **6**, which hardly occurred in the calli. The productivity of the liquid cultures was not related to either the productivity of the original calli or the growth rate of the biomass in the liquid cultures (*165*).

Further progress in the field of tissue culture has resulted from the use of *Agrobacterium tumefasciens*, responsible for crown gall disease, and *A. rhizogenes*, which induces hairy root disease in dicotyledonous plants. Crown gall callus of *Tagetes* transformed with *A. tumefasciens* grew indefinitely in the absence of hormones and never regenerated shoots or roots (*82*). While such undifferentiated *T. erecta* had low thiophene levels, transformation of *T. minuta* did not adversely affect the level of these compounds (*82*). The variables controlling the production of polythiophenes in cultures of *T. patula* transformed with *A. tumefasciens* are quite complex. Infection of *T. patula* resulted in the formation of rhizoid tumors which, when subcultured in liquid medium, produced vascular bundles similar to those of untransformed roots (*121*). These transformed "root" cultures were subcultured over long periods of time (longer than one year), during which period the synthesis and secretion rate of thiophenes were constant. Most importantly, extracts from non-transformed cell suspensions showed the **7**-acetate esterase activity identified by Sütfeld and Towers in intact *T. patula* plants. No such studies on transformed cells have yet been reported.

Hairy root cultures of *T. patula* transformed with *A. rhizogenes* were also established. They grew rapidly and were stable over many months in subsequent cultures. Two diacetylenes were identified, the primary alcohol **6** and its acetate **7** (*105*), which were synthesized regularly in all cultures. This contrasts with the behavior of crown gall callus cultures (*192*). Most surprisingly, exposure of these hairy roots to light induced greening, and even more rapid growth.

The potential of such hairy root cultures in biosynthetic studies remains to be fully realized, but it is clear that the system invites exciting biosynthetic and enzymatic studies on di- and trithiophenes.

VIII. The Synthesis of Di- and Trithiophenes

The synthesis of bi- and trithiophenes has two major components, the synthesis of the thiophene moiety and the introduction of the proper substituents. Although not necessarily very stable, all naturally occurring bi- and trithiophenes isolated to date are relatively simple structurally in the sense that the substituents consist of only a few carbon atoms and contain no more than one asymmetric center. Interestingly, the absolute configuration of such compounds containing a chiral center has in no instance been reported. Although no longer a great challenge, synthesis of these compounds is nonetheless extremely important when the timing, location or scale of an investigation would prohibit the collection of plants and their extraction for known constituents.

The main approaches will be illustrated here, without an attempt at an exhaustive compilation of all syntheses reported to date.

VIII. 1. The Thiophene Rings

The heterocyclic ring can be most readily synthesized by introducing sulfur into a four-carbon atom fragment using one of the two methods outlined in Scheme 10. The first reaction is actually carried out by reacting the 1,3-butadiyne with Na_2S in ethanol (*216*); the second is performed by treating the 1,4-diketone with H_2S in acidic medium or with P_4S_{10} (*210*) or an equivalent, such as Lawesson's reagent. These reactions are versatile in the sense that either or both R_1 and R_2 may themselves be thiophene rings, thus leading to bi- or trithiophene derivatives.

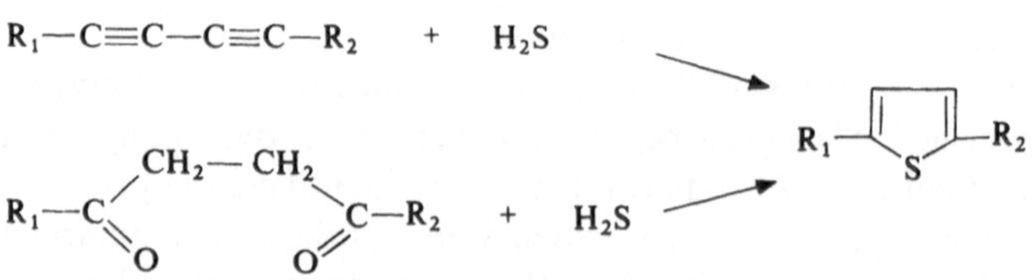

$$R_1-C\equiv C-C\equiv C-R_2 \quad + \quad H_2S$$

$$R_1-C(=O)-CH_2-CH_2-C(=O)-R_2 \quad + \quad H_2S$$

$\longrightarrow$ 2-R_1-5-R_2-thiophene

Scheme 10. The synthesis of thiophenes from acyclic precursors

The starting materials for these syntheses are easily accessible. Standard coupling of terminal acetylenes with $CuCl_2$ in the presence of air (Glaser procedure) produces symmetrical 1,3-butadiynes. Unsymmetrical ones are accessible through the Cadiot-Chodkiewicz coupling of a terminal acetylene with a bromoacetylene, also in the presence of copper salts (*77*). Both kinds of butadiynes can also be synthesized through organoborane chemistry by adding the two different 1-lithioacetylenes successively to B-methoxy-9-BBN and coupling them in the oxidative decomposition step of the ate complex (*218*) (Scheme 11).

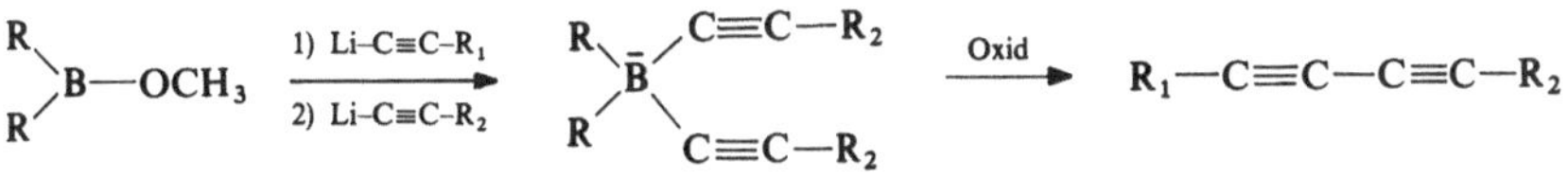

$$R_2B{-}OCH_3 \xrightarrow[2)\ Li{-}C{\equiv}C{-}R_2]{1)\ Li{-}C{\equiv}C{-}R_1} R_2\bar{B}(C{\equiv}C{-}R_2)(C{\equiv}C{-}R_2) \xrightarrow{Oxid} R_1{-}C{\equiv}C{-}C{\equiv}C{-}R_2$$

Scheme 11. 1,3-butadiynes via organoboranes

The conversion of a 1,4-diketone into a thiophene ring is a very convenient method, but not when extremely pure products are needed, for example when biological or physical properties are to be measured. Furans are formed competitively with thiophenes; the extent of their formation apparently depends on the substitution pattern and probably on reaction and work-up conditions as well. Occasionally, the separation of these two products is extremely difficult (*260*).

Symmetrical 1,4-diketones attached to thiophene rings are conveniently formed by direct oxidation of the enolate ion of a methyl ketone with Cu_2Cl_2 and air (*146*) or with iodine (*167*). A more tedious approach involves oxidation of a trimethylsilyl enol ether derivative with AgO (*8*), with Ag_2O (*134*) or with C_6H_5IO in the presence of BF_3 (*187*). The synthesis of **T1**, for example, can start with the readily available 2-acetylthiophene (Scheme 12).

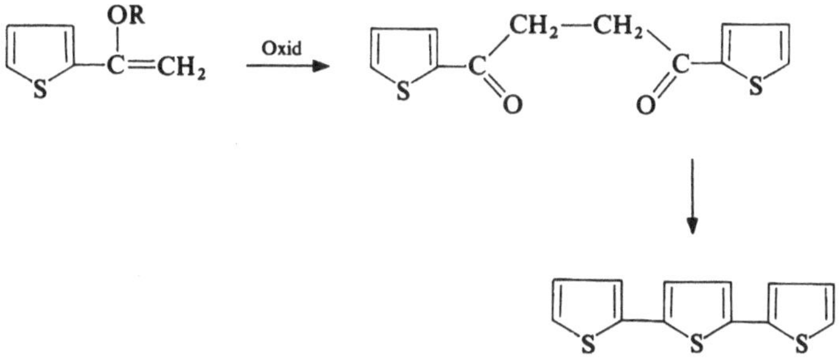

Scheme 12. Synthesis of **T1** from 2-acetylthiophene

Unsymmetrical 1,4-diketones are less readily available, but KOOREMAN and WYNBERG have described the synthesis of a number of substituted trithiophenes by cyclization of unsymmetrical 1,4-diketones secured through a Stevens rearrangement as shown in Scheme 13 (*167*).

$$R_1COCH_2X + R_2COCH_2N(CH_3)_2 \longrightarrow R_1COCH_2\text{—}\overset{+}{N}(CH_3)_2(CH_2COR_2) \xrightarrow{\text{Base}}$$

$$\left[R_1COCH(CH_2COR_2)\text{—}N(CH_3)_2\right] \xrightarrow{\text{Acid}} R_1COCH{=}CHCOR_2 \xrightarrow[\text{Acid}]{SnCl_2} R_1COCH_2CH_2COR_2$$

Scheme 13. Synthesis of unsymmetrical 1,4-dicarbonyl compounds

Another approach to 1,4-dicarbonyl products is also quite versatile and powerful. It involves the conjugate addition of a carbonyl anion equivalent derived from an aldehyde to an α,β-unsaturated carbonyl compound, using either cyanide or thiazolium ion catalysis (Stetter reaction). In the scheme shown (Scheme 14), it is also possible to have $R_3 = H$, by using vinyl ketones or their Mannich base equivalents (*225*).

$$R_1CHO + R_3CH{=}CHCOR_2 \longrightarrow R_1COCH(R_3)CH_2COR_2$$

Scheme 14. Synthesis of 1,4-dicarbonyl compounds by the Stetter reaction

In a third approach to thiophenes, six-membered 1,2- or 1,4-dithiin rings are synthesized (*190*) and subsequently decomposed with formal loss of one sulfur atom (Scheme 15).

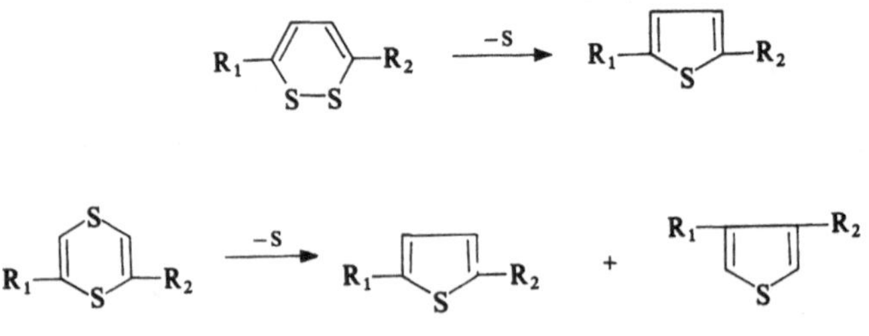

Scheme 15. Thiophenes from 1,2- and 1,4-dithiins

The synthesis of 1,2-dithiins, particularly those with sensitive substituents, is not as simple as that of thiophenes themselves by the previous methods and is no better for preparative work. Synthesizing 1,4-dithiins is much easier than preparation of the 1,2-isomers, but unavoidably leads to a mixture of isomers. This approach is more valuable for the synthesis of 3,4 or 2,4-disubstituted thiophenes, which are more difficult to make than 2- or 2,5-substituted thiophenes.

Finally, a modified Wittig reaction has produced a clever synthesis of thiophene rings (*20*) (Scheme 16).

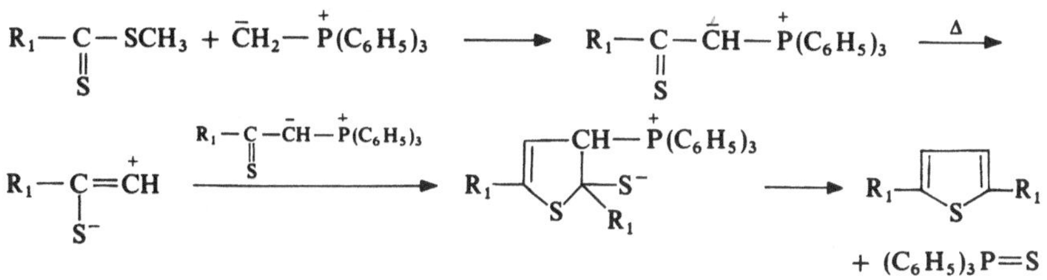

$$R_1-\underset{\underset{S}{\parallel}}{C}-SCH_3 + \bar{C}H_2-\overset{+}{P}(C_6H_5)_3 \longrightarrow R_1-\underset{\underset{S}{\parallel}}{C}-\bar{C}H-\overset{+}{P}(C_6H_5)_3 \xrightarrow{\Delta}$$

$$R_1-\underset{\underset{S^-}{|}}{C}=\overset{+}{C}H \xrightarrow{R_1-\underset{\underset{S}{\parallel}}{C}-\bar{C}H-\overset{+}{P}(C_6H_5)_3} \longrightarrow + (C_6H_5)_3P=S$$

Scheme 16. BESTMANN's synthesis of thiophenes

VIII. 2. Connecting the Thiophene Rings

Coupling reactions are powerful methods for creating polythiophenes. Ulmann coupling of 2-iodothiophene with copper was first used for making 2,2′-bithiophene itself. Although this reaction is not clean, as it also produces **T1** and higher oligomers, it has the advantage of involving simple and relatively inexpensive materials. However, the coupling reaction of 2-lithiothiophenes by oxidation with $CuCl_2$ in dimethylformamide is an excellent method for making 2,2′-bithiophene and for selectively making other even-numbered thiophene oligomers (*147*).

Thiophene rings can be coupled through organoboranes (*148*). The key step is the oxidative treatment of an ate complex, itself formed by

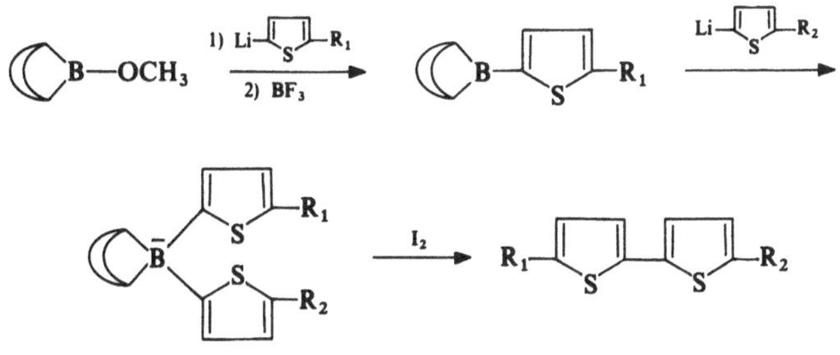

Scheme 17. Bithiophenes via organoboranes

addition of an organolithium compound to a trisubstituted borane (Scheme 17). Two thiophene rings must be present in the ate complex, but the order of their introduction might differ from that shown below, at least in principle, if other borane precursors were chosen. Borinic acid derivatives have also been used for thiophene coupling reactions, but the yields are not very good (*91*).

A photochemical coupling reaction of 2-halothiophenes recently published was used for the preparation of a number of naturally occurring bithiophenes. It involves the photolysis of a 2-iodothiophene with another thiophene molecule (*87*, *88*, *89*, *90*). This method was used

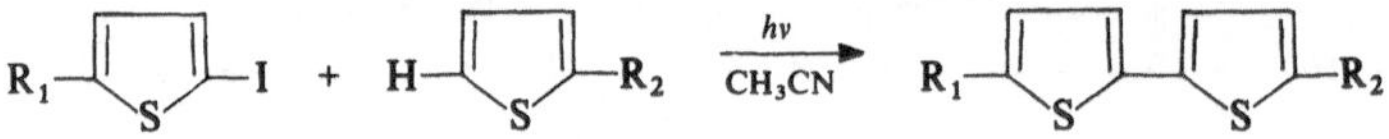

Scheme 18. Bithiophenes *via* photochemical coupling

for the conversion of 5-iodo-2-thiophenecarboxaldehyde into **1** (*90*), **16** (*87*), **28**, **29**, **32**, **42** (*88*) and **31** (*88*, *89*) but does not seem to have been applied to the synthesis of trithiophenes.

A third approach involves the Kumada procedure, in which the Grignard reagent of a thiophene derivative is coupled with a halothiophene (Scheme 19). It is probably the most reliable method for making

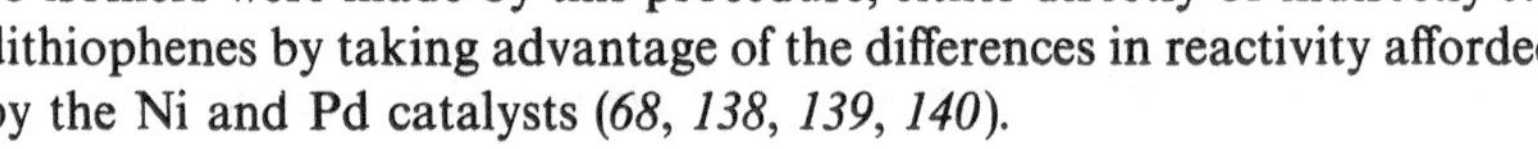

Scheme 19. Bithiophenes via Kumada coupling

polythiophenes in excellent yield. Recently, for example, **T1** as well as its 13 isomers were made by this procedure, either directly or indirectly *via* dithiophenes by taking advantage of the differences in reactivity afforded by the Ni and Pd catalysts (*68*, *138*, *139*, *140*).

VIII. 3. Introduction of the Side Chains

As suggested above, the synthesis of many substituted bi- or trithiophenes can be carried out by using precursors bearing the substituents needed in the final products. Alternatively, one or two substituents may be introduced or modified after the ring system is in place. Almost all naturally occurring compounds of interest are substituted at the α-positions of the thiophene rings, the positions most readily attacked in electrophilic reactions (for example in halogenation, Friedel-Crafts or

Vilsmeier reactions), or in reactions with nonaqueous bases (such as n-BuLi). Consequently, the main challenges in these syntheses do not emanate from the regioselectivity in the substitution steps.

Many syntheses have introduced substituents onto the thiophene ring by reacting a 2-iodothiophene with an organocopper derivative. One such example (*11*) is shown here in Scheme 20. In the same paper synthesis of the above reaction product was also accomplished by reacting ethylene oxide with the Grignard reagent from 2-ethynylbithiophene.

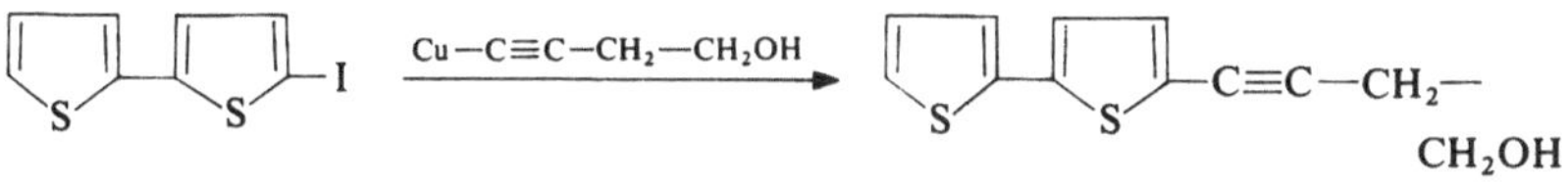

$$\text{(bithienyl)}-I \xrightarrow{Cu-C\equiv C-CH_2-CH_2OH} \text{(bithienyl)}-C\equiv C-CH_2-CH_2OH$$

Scheme 20. Coupling of an iodothiophene with an organocopper derivative

The aldehyde group can be readily introduced at an α-position of a mono-, bi-, or trithiophene by the traditional Vilsmeier reaction with $POCl_3$ and $HCON(CH_3)_2$ or by reacting the corresponding α-lithiothiophene with $HCON(CH_3)_2$ (*223*). It provides a source for many of the common substituents. Thus carboxylic acids may be prepared by oxidation, alcohols and their esters by reduction and further esterification, vinyl groups by a Wittig reaction, etc. Of course, carboxylic acid groups can be introduced by carboxylation of a thienyl Grignard or lithium derivative (*149*) or by carbonylation in the presence of a Pd complex as catalyst (*206*). Thiophenecarboxylic acids may be unusually insoluble in base. This was demonstrated, for example, with α-terthienylcarboxylic acid (*149*). This property perhaps explains why no carboxylic acids have been isolated from natural sources in the trithiophene series, and only three as minor products in the dithiophene series.

As many products of interest bear acetylenic side chains, some of the techniques used for introducing these substituents are outlined in the equations below (Scheme 21). Although the introduction of an acetylenic group by reaction of PCl_5 with an acetyl substituent, followed by treatment with $NaNH_2$, has been reported (*36*), the method does not work well with 2-ethynylthiophene itself (*66*).

The acetylenic group can be a handle for further functionalization. Typical examples are conversion into a propyne, as in the reaction of a sodium salt with methyl iodide (*223*), or into the 3-butene-1-yne substituent found in **1**. The latter reaction, illustrated in Scheme 22, forms product **1** by thermal dehydration of the alcohol obtained when the Grignard reagent reacts with acetaldehyde (*199*). In an alternative

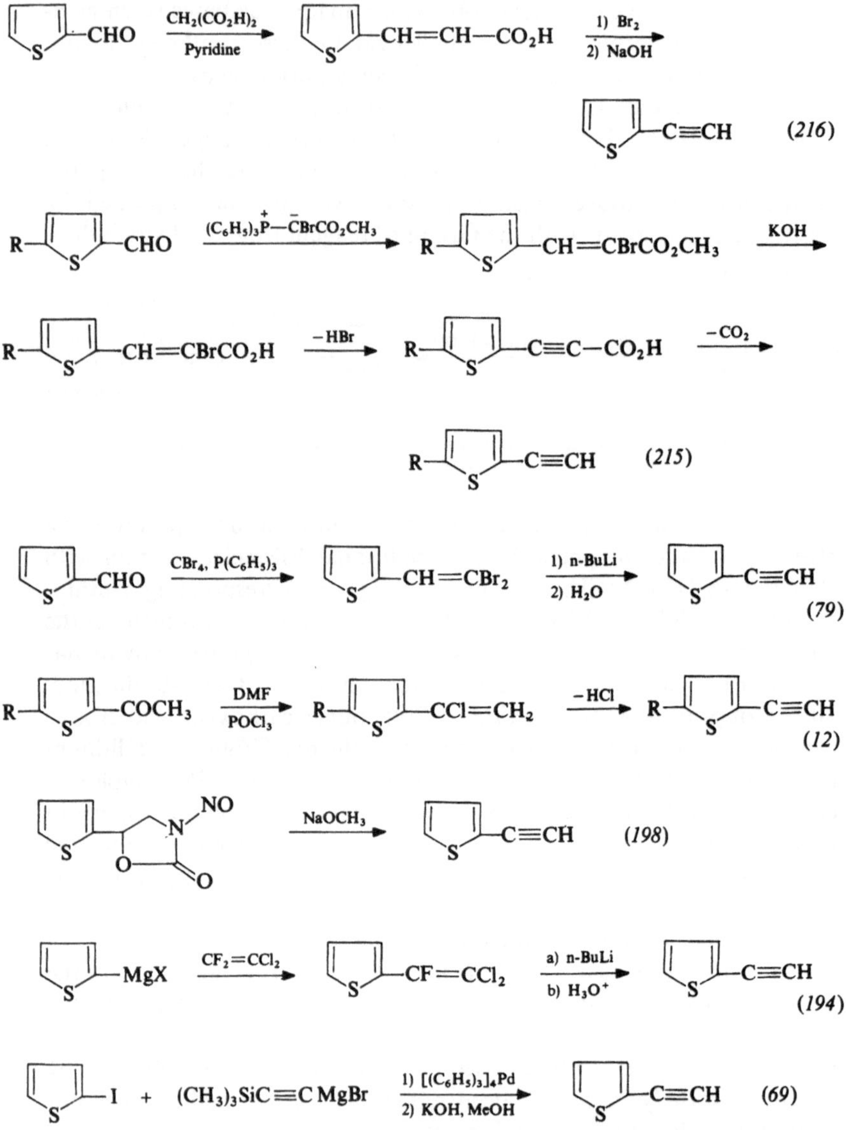

S —CHO $\xrightarrow[\text{Pyridine}]{CH_2(CO_2H)_2}$ S —CH=CH—CO_2H $\xrightarrow[\text{2) NaOH}]{\text{1) }Br_2}$ S —C≡CH (*216*)

R— S —CHO $\xrightarrow{(C_6H_5)_3\overset{+}{P}—\overset{-}{C}BrCO_2CH_3}$ R— S —CH=$CBrCO_2CH_3$ $\xrightarrow{KOH}$

R— S —CH=$CBrCO_2H$ $\xrightarrow{-HBr}$ R— S —C≡C—CO_2H $\xrightarrow{-CO_2}$ R— S —C≡CH (*215*)

S —CHO $\xrightarrow{CBr_4,\ P(C_6H_5)_3}$ S —CH=CBr_2 $\xrightarrow[\text{2) }H_2O]{\text{1) n-BuLi}}$ S —C≡CH (*79*)

R— S —$COCH_3$ $\xrightarrow[POCl_3]{DMF}$ R— S —CCl=CH_2 $\xrightarrow{-HCl}$ R— S —C≡CH (*12*)

S —(N-NO, O, =O oxazolidinone) $\xrightarrow{NaOCH_3}$ S —C≡CH (*198*)

S —MgX $\xrightarrow{CF_2=CCl_2}$ S —CF=CCl_2 $\xrightarrow[\text{b) }H_3O^+]{\text{a) n-BuLi}}$ S —C≡CH (*194*)

S —I + $(CH_3)_3SiC≡C\,MgBr$ $\xrightarrow[\text{2) KOH, MeOH}]{\text{1) }[(C_6H_5)_3]_4Pd}$ S —C≡CH (*69*)

Scheme 21. Representative syntheses of 2-ethynylthiophenes

approach, the primary alcohol is dehydrated by treatment of the corresponding tosylate with KOH (*11*). An almost quantitative synthesis of the same product was achieved by coupling the Grignard reagent with vinyl bromide, in the presence of a Pd complex catalyst (*205*, *206*) as shown in

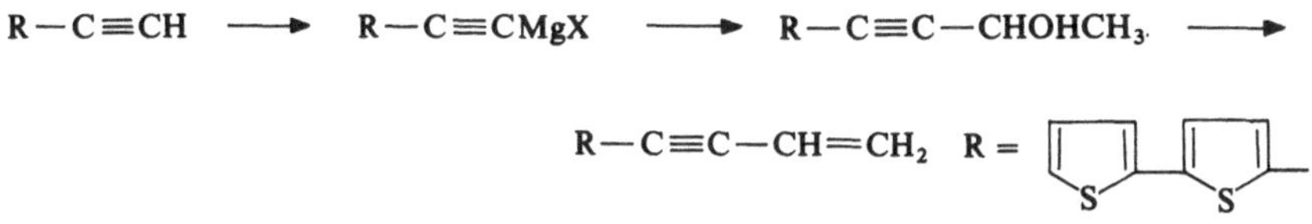

$R-C\equiv CH \longrightarrow R-C\equiv CMgX \longrightarrow R-C\equiv C-CHOHCH_3 \longrightarrow$

$R-C\equiv C-CH=CH_2$ R =

Scheme 22. Synthesis of **1** via an acetylenic Grignard reagent

$-C\equiv CMgBr \xrightarrow[Pd(PPh_3)_4]{BrCH_2=CH_2} -C\equiv C-CH=CH_2$

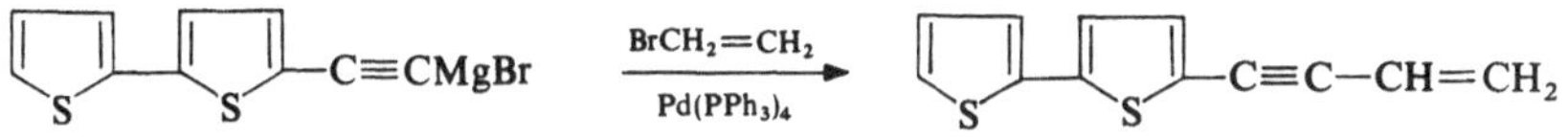

Scheme 23. Pd-catalyzed coupling of a Grignard reagent with a vinyl halide

Scheme 23. The coupling may be carried out instead with the acetylenic compound itself, using the same catalyst in the presence of CuI and base under phase-transfer conditions (*206*) (Scheme 24). Of course, the eneyne

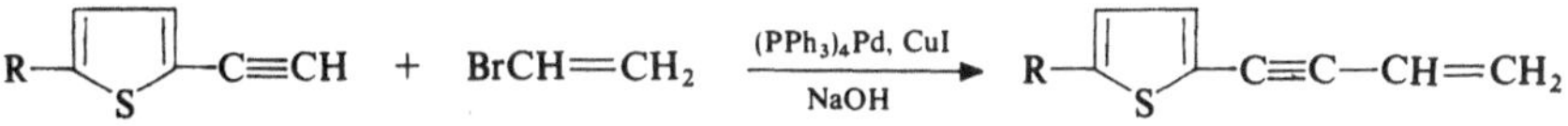

$R-\!-C\equiv CH + BrCH=CH_2 \xrightarrow[NaOH]{(PPh_3)_4Pd,\ CuI} R-\!-C\equiv C-CH=CH_2$

Scheme 24. Pd-catalyzed coupling of an acetylene with a vinyl halide

side chain could also be introduced in one unit as shown in Scheme 25 (*13*).

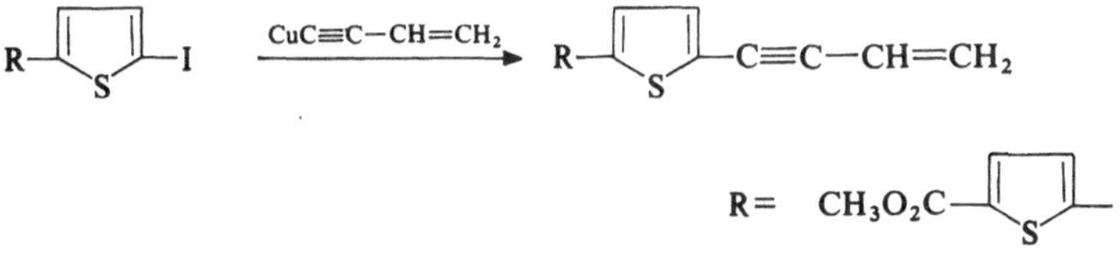

$R-\!-I \xrightarrow{CuC\equiv C-CH=CH_2} R-\!-C\equiv C-CH=CH_2$

R= CH_3O_2C-

Scheme 25. Introduction of the eneyne substituent

It is likely that other metal acetylides, such as tin derivatives (*207*) could be used in addition to the more common magnesium and copper compounds.

It should be noted that the common 1-propynyl substituent does not have to be introduced by methylation of an ethynylthiophene precursor. It can be introduced either through coupling of the acetylenic Grignard reagent with a 2-halothiophene in the presence of a Pd complex, or through decomposition of an heterocyclic precursor (*253*) (Scheme 26).

Subsequent functionalization of the other α-position has led to some of the components of *Arctium lappa* as shown in Scheme 27 (*253*).

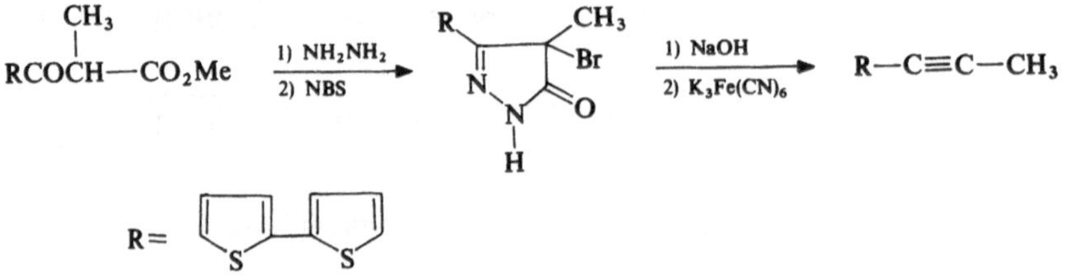

$RCOCH(CH_3)-CO_2Me$ → 1) NH_2NH_2, 2) NBS → [pyrazolone with CH_3, Br] → 1) NaOH, 2) $K_3Fe(CN)_6$ → $R-C{\equiv}C-CH_3$

R =

Scheme 26. Synthesis of a 1-propyne side chain

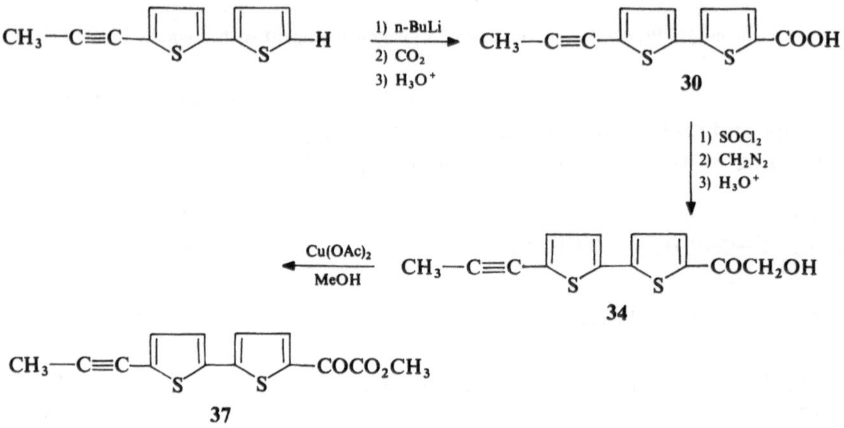

$CH_3-C{\equiv}C-$ ... $-H$ → 1) n-BuLi, 2) CO_2, 3) H_3O^+ → $CH_3-C{\equiv}C-$ ... $-COOH$ **30**

1) $SOCl_2$, 2) CH_2N_2, 3) H_3O^+

$CH_3-C{\equiv}C-$ ... $-COCH_2OH$ **34** → $Cu(OAc)_2$, MeOH

$CH_3-C{\equiv}C-$ ... $-COCO_2CH_3$ **37**

Scheme 27. Synthesis of components of *Arctium lappa*

The synthesis of compound **61** bearing one substituent at a β-position has been accomplished by coupling an iodothiophene bearing the 1-propynyl substituent with 2-ethynyl-3-acetoxythiophene (*52*). The latter

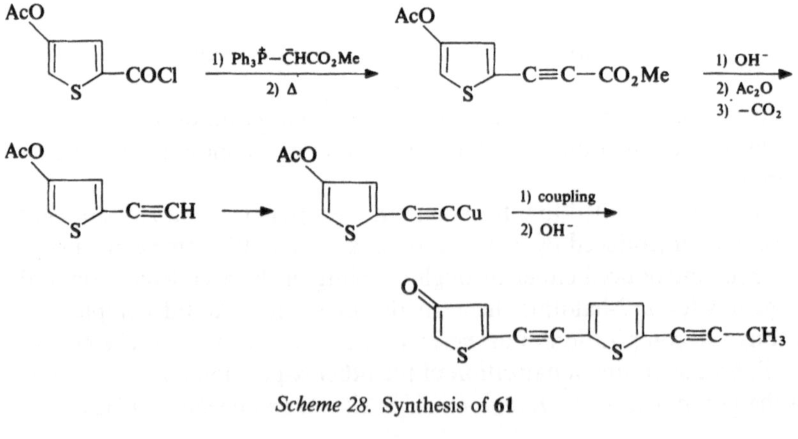

AcO ... COCl → 1) $Ph_3\overset{+}{P}-\overset{-}{C}HCO_2Me$, 2) Δ → AcO ... $C{\equiv}C-CO_2Me$ → 1) OH^-, 2) Ac_2O, 3) $-CO_2$

AcO ... $C{\equiv}CH$ → AcO ... $C{\equiv}CCu$ → 1) coupling, 2) OH^-

O ... $C{\equiv}C-$ ... $-C{\equiv}C-CH_3$

Scheme 28. Synthesis of **61**

was prepared by a modified Wittig reaction with an acid chloride, directly producing an acetylene substituent (Scheme 28).

The 1,2-dithienylacetylene moiety has also been synthesized by a Wittig reaction with thiophenecarboxaldehyde and the triphenylphosphine ylide made from a 2-halomethylthiophene, the resulting olefin being successively brominated and dehydrobrominated (Scheme 29) (*36*).

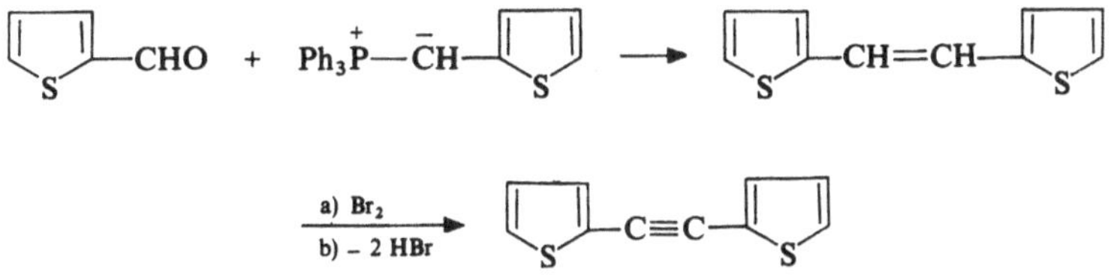

$$\text{C}_4\text{H}_3\text{S}{-}\text{CHO} + \text{Ph}_3\overset{+}{\text{P}}{-}\overset{-}{\text{C}}\text{H}{-}\text{C}_4\text{H}_3\text{S} \longrightarrow \text{C}_4\text{H}_3\text{S}{-}\text{CH}{=}\text{CH}{-}\text{C}_4\text{H}_3\text{S} \xrightarrow[\text{b) } -\,2\,\text{HBr}]{\text{a) Br}_2} \text{C}_4\text{H}_3\text{S}{-}\text{C}{\equiv}\text{C}{-}\text{C}_4\text{H}_3\text{S}$$

Scheme 29. Synthesis of a 1,2-di(2-thienyl)acetylene via an olefin

More expeditiously, the TMS derivative of an ethynylthiophene can be coupled with an iodothiophene using a Pd catalyst (Scheme 30) (*206*). A similar type of reaction is used to couple 2- and 3-halothiophenes with other unsaturated reagents (*231*).

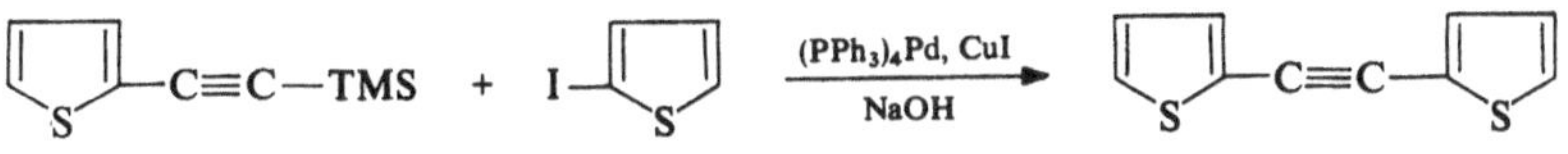

$$\text{C}_4\text{H}_3\text{S}{-}\text{C}{\equiv}\text{C}{-}\text{TMS} + \text{I}{-}\text{C}_4\text{H}_3\text{S} \xrightarrow[\text{NaOH}]{(\text{PPh}_3)_4\text{Pd, CuI}} \text{C}_4\text{H}_3\text{S}{-}\text{C}{\equiv}\text{C}{-}\text{C}_4\text{H}_3\text{S}$$

Scheme 30. Syntheses of 1,2-di(2-thienyl)acetylene via an acetylene

In conclusion, this survey shows that good methods are available for synthesizing all the di- and trithiophene molecules isolated to date. The same methods should be useful for preparing analogs and derivatives.

IX. Photophysical Studies

IX. 1. Background

The establishment of structure-activity relationships has been the ultimate goal of many syntheses of bithiophene and trithiophene derivatives. However, since the biological activity of all the bi- and trithiophenes analyzed to date is either light-dependent or can be traced to the formation of electronically excited molecules, any real understanding of the mechanism(s) of phototoxicity at the molecular level must also take into account the photophysical properties of these molecules.

All photochemical reactions start with the absorption of one photon by one molecule. The energy of the molecule is raised considerably by this process, the exact amount depending upon the wavelength of the light used. For example, photons at 254 nm (produced by a low pressure mercury vapor lamp) correspond to about 113 kcal/mol, a value higher than the bond energy of many bonds in organic molecules. The photon energy values are 82, 71.5, and 36 kcal/mol at 350, 400, and 800 nm respectively; these wavelengths correspond to the maximum emission of commercial near-ultraviolet lamps (UVA lamps), and to the limits of the visible spectrum respectively. In order to successfully interact with a molecule and excite it electronically, one photon must possess exactly the energy corresponding to the difference in energies between the highest occupied molecular orbital and the next higher available orbital, which is the lowest unoccupied molecular orbital of the sensitizer. According to quantum mechanics, the initial excitation proceeds with conservation of spin: a molecule in its singlet ground state produces a short-lived singlet electronically excited state.

$$\text{A(Singlet}_0) + \text{photon} \longrightarrow \text{A (Singlet}_1)$$

Intersystem crossing is the process by which a molecule changes its spin state, for example from an excited singlet to a lower energy triplet, or from a triplet to a lower energy singlet, as exemplified by return of the excited molecule A to ground state.

$$\text{A (Singlet}_1) \longrightarrow \text{A (Triplet}_1) \quad \text{(intersystem crossing)}$$

$$\text{A (Triplet}_1) \longrightarrow \text{A (Singlet}_0) \quad \text{(intersystem crossing)}$$

An excited molecule in a triplet state has a much longer lifetime than in its corresponding singlet. This explains why many reactions involving bimolecular energy transfer occur from the excited triplet state of molecules called sensitizers. The excited sensitizer returns to its singlet ground state while producing the triplet excited state of the quencher, which can then undergo chemical transformations through this indirect transfer of energy from photons (Scheme 31). Note that in this reaction sequence, the quencher is not directly excited electronically by a photon; consequently, its singlet excited state is not involved.

$$\text{Sensitizer (S}_0) + \text{light} \longrightarrow \text{Sensitizer (S}_1) \longrightarrow \text{Sensitizer (T}_1)$$

$$\text{Sensitizer (T}_1) + \text{Quencher (S}_0) \longrightarrow \text{Sensitizer (S}_0) + \text{Quencher (T}_1)$$

$$\text{Quencher (T}_1) \longrightarrow \text{Reaction products}$$

Scheme 31. Photosensitized formation of a triplet excited molecule

A special case is energy transfer to oxygen because this molecule has a triplet ground state. Photosensitized reactions involving oxygen are called *photodynamic reactions.* Here the excited sensitizer returns to its ground state while forming oxygen in its singlet excited state (singlet oxygen is usually noted as $^1O_2$). This type of reaction is also called a photodynamic reaction of Type II.

$$\text{Sensitizer } (T_1) + O_2(T_0) \longrightarrow \text{Sensitizer } (S_0) + O_2(S_1)$$

Chemical quenching is also known. A photodynamic reaction of Type I, for example, is a process in which a molecule in its triplet electronically excited state eventually transfers one electron to O_2, producing superoxide anion radical $O_2^{-\cdot}$ and a radical-cation derived from the sensitizer.

$$\text{Sensitizer}^* + O_2 \longrightarrow \text{Sensitizer}^{+\cdot} + O_2^{-\cdot}$$

Fundamental investigations of light-dependent transformations must study both the excited states and their reactivity. Often, these studies are conducted in the absence of oxygen in order to prevent quenching. However, since most, if not all, light-dependent toxic reactions of bi- and trithiophenes require the presence of oxygen, it is also particularly desirable to investigate the fate of these electronically excited molecules in the presence of oxygen.

IX. 2. Nature of the Excited States

Only one paper has discussed the application of magnetic circular dichroism (MCD) to a study of the excited states of bi- and terthiophenes. The MCD spectra of **T1** and **1** showed a broad asymmetric B term absorption corresponding to the absorption band, and suggested $S_0 \rightarrow S_1$ transitions, as expected from theory. In contrast, quantum-mechanically forbidden $S_0 \rightarrow T_1$ transitions were detected by the same authors in the irradiation of several naturally occurring polyacetylenes (*182*).

Laser flash photolysis is a good technique for directly investigating the excited states of molecules and their reactivity. Unfortunately, the technique has never been applied to 2,2′-dithiophenes. The main focus of all publications in this area to date has been on α-terthienyl and some of its derivatives and analogs, **T1** being the only molecule common to all the papers.

IX. 2. 1. Singlet Excited States

The unusually intense fluorescence of trithiophene derivatives has led to one patent on the use of such compounds as optical brighteners of fabrics (*10*). The lowest excited state of **T1** and three derivatives must have π–π^* character, as their absorption maxima and fluorescence quantum yields are very similar in solvent systems of very different polarity (*204*). Oxygen does not quench the singlet excited state of **T1** (*209*), which has a lifetime shorter than 1×10^{-9}s (*104*) and a quantum yield of fluorescence of *ca.* 0.06–0.08 (*204, 211*). A more complete analysis of the fluorescence of **T1** has been carried out in n-decane down to 4.2 K (*21*).

IX. 2. 2. Triplet Excited States and Singlet Oxygen Formation

Following the demonstration that the sensitizer α-terthienyl could generate singlet oxygen (*14*), the first quantitative studies attempted to compare the yield of singlet oxygen produced by different sensitizers under standard conditions (*92, 183*), and to study the photooxidation of cholesterol to its 5-α derivative (*107*).

Whereas the fluorescence of **T1** and its congeners is very conspicuous, their phosphorescence could not be directly detected in common solvents. Studies with quenchers of known triplet energy placed that of **T1** between 42 and 47 kcal/mol (*104*). In recent work, however, SCAIANO *et al.* cleverly overcame the lack of phosphorescence by directly measuring the very weak $S_0 \rightarrow T_1$ absorption in heavy-atom solvents, where significant enhancement is observed. The triplet energy of α-terthienyl was thus determined to be 39.7 ± 1.5 kcal/mol (*208*).

An excellent photosensitizer is characterized by a high quantum yield for intersystem crossing, Φ_{ISC}, and a long lifetime of the triplet excited state. In any given compound these properties may be strongly affected by the environment. In 95% ethanol, the values of Φ_{ISC} for several trithiophenes were found to be between 0.2 and 0.35 (*204*). The highest value (0.35) was for the 5,5″-dibromo derivative due to the heavy atom effect of the substituents. FOOTE had reported a quantum yield of 0.8 for α-terthienyl in benzene (*106*). The value for Φ_{ISC} must be the upper limit of the quantum yield for $^1O_2$ formation under the same conditions. The quantum yield of bleaching of 2,5-diphenylisobenzofuran by $^1O_2$ agreed well with Φ_{ISC} (*204*). A later investigation of α-terthienyl gave essentially the same values for the quantum yields of intersystem crossing and $^1O_2$ formation in ethanol (*150*). Thus, the results obtained by two different methods were in agreement, while a more recent study by SCAIANO *et al.* (*211*), using laser-induced optoacoustic calorimetric techniques, led to higher values for the quantum yields of intersystem crossing (*ca.* 1) as well as for that of singlet oxygen formation, the latter being almost the same (0.7–0.75) in all the solvents tested. Further experimental studies will undoubtedly seek the origin of the discrepancies between the results obtained with these different techniques. In his most recent work, SCAIANO also compared the singlet oxygen emission produced by α-terthienyl with that of reference compounds. He reported quantum yields ranging from 0.67 (in $CHCl_3$) to 0.75 (in C_6H_6 and CH_2Cl_2), with intermediate values in CH_3CN (0.68), CD_3OD (0.67, 0.70), and $CDCl_3$ (0.73) (*211*). All derivatives of α-terthienyl tested had very similar $^1O_2$ generating efficiencies in $CDCl_3$, ranging between 0.69 and 0.93. A similar solvent effect appears to operate in the dibromo derivative, which had $\Phi_{ISC} = 0.35$ in ethanol, but $\Phi_{ISC} = 0.69$ in $CDCl_3$. The transient triplet state absorption spectra were directly observed, and no phosphorescence could be detected. In all cases, the sum of the quantum yields of luminescence and intersystem crossing was less than 1, indicating efficient internal conversion processes. The concentration dependence of the lifetime of the excited sensitizers was explained by ground state molecules quenching the triplet states. This interpretation was supported by the long lifetime observed in micelles which contained, on the average, one molecule of α-terthienyl per micelle, where bimolecular quenching is unlikely to occur (*204*).

In the work dealing with α-terthienyl in 95% ethanol, the experimental values for the quantum yields of singlet oxygen formation and intersystem crossing obtained by two different experimental techniques concorded. By contrast, the different values obtained in the other solvents, were obtained by a single technique, and unfortunately the experimental results did not include values in 95% ethanol. Thus, a critical evaluation of the results obtained by different techniques in the several laboratories is not yet possible.

IX. 2. 3. Photoinduced Electron Transfer Reactions

REYFTMANN *et al.* could not detect spectroscopically any solvated electrons when α-terthienyl was irradiated in alcohol or aqueous media (*204*). When oxygen was added, therefore, no superoxide ion was detected. It would have been formed by initial photoionization and electron transfer to O_2 via these solvents. Although slow quenching was observed in the presence of $(C_2H_5)_3N$, an electron donor, it was the quenching of the triplet state of the sensitizers which was monitored rather than the spectrum of the expected $(C_2H_5)_3N^{+\cdot}$. This result does not prove that excited triplet states of the trithiophenes are electron acceptors. Physical quenching or other modes of chemical interaction might have been involved, and physical quenching of the triplet states of α-terthienyl and its 5-iodo and 5,5″-diiodo derivatives by anthracene and β-carotene was actually observed. By contrast, electron transfer reactions from α-terthienyl and its derivatives to an acceptor were observed in methanol by EVANS *et al.* (*104*): the excited states of the sensitizers reduced methyl viologen (MV^{2+}) to $MV^{+\cdot}$, which was observed spectroscopically, but the back electron transfer reactions were also very efficient (90–100%). SCAIANO *et al.* (*209*) also observed photoinduced electron transfer from α-terthienyl to tetracyanoethylene in acetonitrile and characterized [α-terthienyl]$^{+\cdot}$ whose absorption maximum is at 530 nm in CH_3CN. The fate of the pair of radicals created by electron transfer from **T1** and analogs to electron acceptors has been carefully investigated in micelles under anaerobic conditions (*103*). The rate of decay was markedly decreased by an applied magnetic field.

Often, a photosensitizer can be involved in photodynamic reactions of both Type I (electron transfer) and Type II (singlet oxygen formation). Obviously, a sensitizer generating singlet oxygen with a high quantum yield can only generate superoxide with a low quantum yield in its primary process. For this reason, it is not surprising that α-terthienyl did not generate superoxide in the organic solvents mentioned above.

The electron transfer reactions from excited α-terthienyl to organic acceptors in organic solvents have their counterpart in aqueous medium, in which the quantum yield of singlet oxygen formation is greatly reduced (*150*). Solvation of the pair of radical-ions, which probably diminishes the ease of a subsequent electron transfer in the reverse direction, best explains the stabilizing effect of water compared with organic solvents.

$$\alpha\text{-terthienyl}^* + O_2 \longrightarrow ([\alpha\text{-terthienyl}]^{+\cdot})_{aq} + (O_2^{-\cdot})_{aq}$$

The efficiency of the electron transfer from α-terthienyl to oxygen was estimated to be lower than 1% (*209*). The efficiency is somewhat higher, 6%, with 2,2′-bithienyl (*103*). The generation of superoxide was revealed experimentally by the photoreduction of cytochrome c or of nitro blue tetrazolium sensitized by α-terthienyl in aqueous solution under aerobic conditions, a reaction which is suppressed in the presence of the very specific enzyme superoxide dismutase. Electron transfer reactions can also occur directly from electronically excited α-terthienyl to cytochrome c, but they are not favored in the presence of oxygen (*150*).

Although the mechanism is not yet fully understood, superoxide ion formation in water (measured by its reaction with cytochrome c) can be greatly enhanced by the presence of a small amount of ethylenediaminetetraacetic acid (EDTA). Since the excited state of α-terthienyl has been observed not to react with EDTA in argon (*18*), other possibilities must be considered. Perhaps metal ion impurities which normally intercept electrons generated from excited α-terthienyl, thus lowering the extent of cytochrome reduction, are chelated by EDTA. Alternatively, $O_2^{-\cdot}$ may be produced when the initially formed $^1O_2$ reacts with EDTA. Thus α-terthienyl may play a catalytic role, whereas in the absence of EDTA the sensitizer disappears *via* chemical reactions of the radical-cation formed. A third possibility is that the first step involves the transfer of one electron from α-terthienyl to O_2, which is followed by a second step in which EDTA transfers one electron to $[\alpha\text{-terthienyl}]^{+\cdot}$, thus regenerating the sensitizer and allowing it to participate in further electron transfers to cytochrome c. The latter mechanism is more probable because reaction of $^1O_2$ independently generated in the gas phase over a mixture of cytochrome c and EDTA did not result in the reduction of the former.

The easy generation of $O_2^{-\cdot}$ observed *in vitro* has not been confirmed with *Escherichia coli* strains *in vivo* (*240*). One probable reason is that the hydrophobicity of α-terthienyl forces the compound to become attached to or imbedded in the organic molecules constituting the biological target. In such a situation, homogeneous reactions in aqueous media are therefore no longer important for the biological activity which is then dominated by singlet oxygen damage to membranes.

IX. 2. 4. Other Electron Transfer Reactions

It is not known whether plants of the family Compositae use endogenous polythiophene molecules to store and/or transport electrons generated through photosynthesis or through metabolic redox reactions. Should they do so, bithiophenes, trithiophenes and higher homologues would fulfil, *in vivo*, functions which have been extensively investigated in the laboratory.

Polythiophenes can be generated from 2,5-disubstituted substrates, for example by reaction with Zn in the presence of Ni or Pd salts (*265*), or with Mg in the presence of either Ni salt (*172*, *212*, *266*) or Z-1,4-dichlorobutene (*125*). Alternatively, they can be formed by electrochemical oxidation of thiophene or small thiophene oligomers, including α-terthienyl (*268*, *269*). The polythiophenes may be obtained as powders or films which, once doped through electrochemical oxidation or reaction with a chemical oxidant such as iodine, conduct electricity. The properties depend critically on a number of parameters associated with the synthetic procedures, and conducting as well as semi-conducting material can be obtained (*248*). Polythiophenes have already been used in a number of practical applications which include the design of electrochromic display materials (*195*) and transducers (*189*), and the manufacture of batteries (*162*). The literature in this area is far too extensive to be detailed in this review.

The final understanding of many physical, photophysical and biological properties of polythiophenes will eventually correlate experimental information with three-dimensional structure. It is noteworthy, therefore that crystallographic data on α-terthienyl have been published (*246*). The molecule is almost planar, and the sulfur atoms on adjacent rings face in opposite directions. However, as always, it is risky to infer structural information on molecules in solution from such crystallographic data.

X. Photochemical Reactions

Upon irradiation, substituted thiophenes undergo rearrangements which are also known to occur in other five-membered heterocyclic rings such as furans, oxazoles and pyrazoles (*170*), in which a substituent appears to migrate from an α- to a β-position. This reaction was discovered in WYNBERG's laboratory, where it was studied extensively (*163*, *258*, *259*, *261*, *262*, *263*). The transformation is due not to the migration of a group from one carbon atom to an adjacent one, but to a ring transformation in which the substituent remains attached to the same carbon atom throughout. Although different mechanisms have been advanced, a $[2_\sigma + 2_\pi]$ electrocyclic process conveniently accounts for the formation of a cyclopropene intermediate, which undergoes ring closure to the isomerized product at the other cyclopropene carbon atom (Scheme 32). The rearrangement is believed to take place from an excited singlet state (*164*). In principle, this reaction should be reversible and therefore lead to a photoequilibrium, but the reverse has seldom been

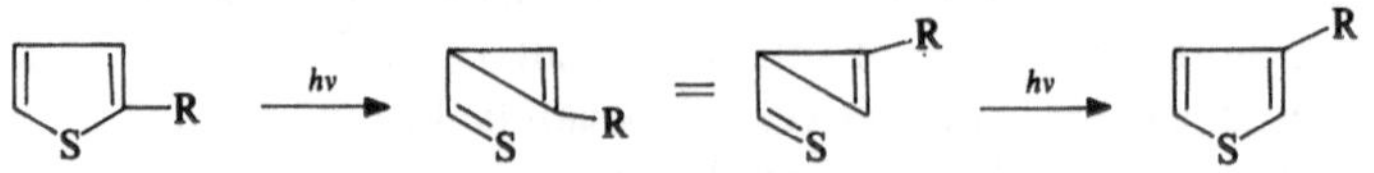

Scheme 32. Photochemical thiophene isomerization

seen (*141, 257*). Perhaps steric reasons (repulsion between the large sulfur atom and the R group) are at play in controlling the ring closure step.

A monosubstituted bithiophene could lead to eight different products upon photolysis, should the typical isomerization occur at all three points of attachment on the rings in all possible manners. The possible number of isomeric products increases rapidly when we consider unsymmetrically disubstituted bithiophenes, trithiophenes, and substituted trithiophenes. Surprisingly, not a single one of these isomers has ever been reported from any of the naturally occurring products covered in this article. Even α-terthienyl, the most thoroughly investigated molecule, has never yielded one well-defined photoproduct. The only compounds that appear to have a photochemical origin are the *cis*- and *trans*-cyclobutanes isolated from *Cardopatium corymbosum* (*217*), which are formally $[2_\pi + 2_\pi]$ cycloaddition dimers of **1** (Scheme 33).

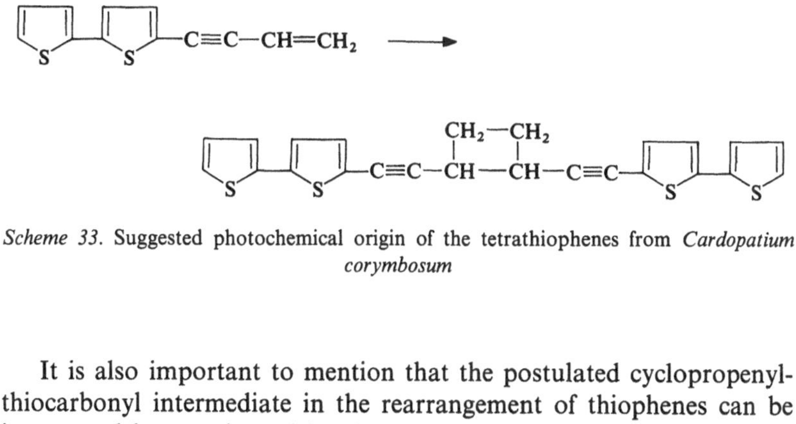

—C≡C—CH=CH$_2$ ⟶

CH_2—CH_2

—C≡C—CH—CH—C≡C—

Scheme 33. Suggested photochemical origin of the tetrathiophenes from *Cardopatium corymbosum*

It is also important to mention that the postulated cyclopropenylthiocarbonyl intermediate in the rearrangement of thiophenes can be intercepted by reaction with primary amines, thus producing pyrroles (*81*) (Scheme 34). Although this reaction has not yet been demonstrated *in vivo*, it could lead to severe modifications of proteins, nucleic acids, and other biological molecules containing primary amine groups.

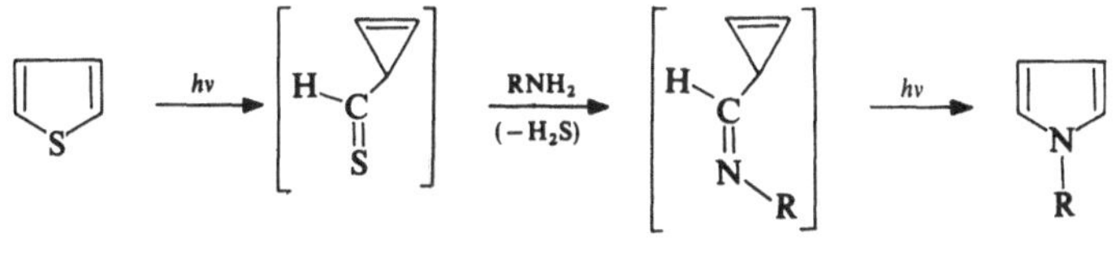

hv; RNH_2 ($-H_2S$); hv

Scheme 34. Photochemical conversion of a thiophene into a pyrrole

XI. Biological Reactions

Most of the studies have been performed with α-terthienyl, and one review dealing with the status of the mechanism of action of α-terthienyl in biological reactions has already appeared (*78*). Here, the activity of di- and trithiophenes will be reviewed according to the biological targets which were investigated.

XI. 1. Effect on DNA

Most of the biological properties of polythiophenes in mammalian cells or cell components have been studied with α-terthienyl as the agent. From the standpoint of immediate safety of handling, it is comforting to know that this substance has a very low level of toxicity in rats and mice (*5*, *201*) and does little damage to DNA. For example, α-terthienyl does not cause the formation of interstrand cross-links in calf thymus DNA (*254*), nor chromosome aberrations nor sister chromatid exchanges in cultured mammalian cells (*173*) when irradiated with UVA. However, it does affect DNA repair in cultured human fibroblasts (*226*). The anaerobic irradiation of calf thymus DNA in the presence of radioactive α-terthienyl led to a slight but measurable level of incorporation of radioactivity into the DNA (*145*). However, no mutations were observed in *Escherichia coli* upon UV irradiation in the presence of α-terthienyl (*239*). Very recently, structural data on DNA damage were obtained when plasmid pBR322 DNA was irradiated in the presence of α-terthienyl: the nicking of supercoiled into relaxed circular DNA was easily detected in this reaction in which one strand of a double stranded circular DNA is cleaved, and the process took place in a time dependent and concentration dependent manner *even in the absence of oxygen* (*250*). Furthermore, aerobic irradiation in the presence of histidine, a good singlet oxygen quencher, led to enhanced DNA damage. Clearly, singlet oxygen reactions cannot be the only important processes in the photobiochemistry of α-terthienyl. Superoxide dismutase, catalase and the antioxidant BHT had no effect on the photosensitized cleavage of pBR322.

XI. 2. Antiviral Properties

Any changes in biological activity resulting from DNA modifications are expected to be readily observed with viruses. HUDSON has extensively investigated the antiviral activity both of naturally occurring thiophenes and of many synthetic analogs (*126*, *128*, *129*, *132*, *237*, *251*). These studies followed the discovery that phenylheptatriyne, a component of many Compositae, modified murine cytomegalovirus (MCMV) in the presence of near-UV radiation. After treatment, the virus penetrated mouse cells normally and the viral DNA entered the nucleus, the normal site of replication. Although the virus particles and the viral DNA retained their integrity, viral genes were not expressed and the virus therefore did not replicate (*131*).

The compounds were tested for activity against two viruses with membranes. They were MCMV, a double stranded DNA herpes virus which replicates in the nucleus of mouse cells, and Sindbis virus (SV), a single-stranded RNA virus which replicates in many types of animal cells. The compounds were also tested on viruses without membranes: phage T_4 and fish infectious pancreatic necrosis virus (IPNV). Antiviral activity was observed only with ultraviolet activation.

α-Terthienyl was one of the most potent antiviral compounds tested, but the activity was virus dependent. In the case of **T1**, for example, the fluence required for decreasing virus infectivity by 99% showed ratios of 1:65:400 for SV, MCMV and T4 viruses respectively (*127*). As with phenylheptatriyne, the MCM virus treated with **T1** in the presence of near-UV penetrated mouse cells normally but failed to replicate (*129*).

Some naturally occurring dithiophenes (**6**, **26**, and **31**) also have antiviral activity. Although less potent than **T1** against SV, they proved more potent against MCMV. Actually, **26** was the most potent of the 31 compounds tested against this virus (*130*). It should be noted, however, that potency in these tests depends on the emission spectrum of the UV light source used since the compounds have different absorption spectra. Consequently, one should correct for the difference in the numbers of photons absorbed when different compounds are compared. Ideally, monochromatic radiation should be used for excitation, the correction factor being deduced from the specific absorbance of each compound at this wavelength. The ranking of the products tested by Hudson would certainly turn out to be quite different with different light sources. This consideration may be important when surveying molecules for eventual medical applications, since available tunable lasers can be used at wavelengths other than those available from standard low pressure mercury UV light sources.

Only time will tell whether viral infections in man, such as AIDS, can be treated successfully with photosensitizing molecules. It is interesting to note, however, that fungal infections in patients with a depressed immune system generally have grave consequences and that drugs with low cytotoxicity but with antiviral and antifungal activity would be quite valuable (all the photoactive antiviral thiophene derivatives tested by HUDSON also have antifungal activity).

HUDSON suggested that two distinct mechanisms of light-dependent activity operated, one involving membranes, the other affecting proteins or nucleic acids when membranes were absent. The possibility that the sensitizer reaches critical locations for damaging DNA and/or viral proteins through specific initial interactions with membranes has not been evaluated.

XI. 3. Biological Effects on Human and Animal Skin

No naturally occurring bi and trithiophenes are currently used as drugs. However, plants containing such compounds have been used in traditional medicine. For example, the juice of *Eclipta alba* leaves has reportedly been used in India for the treatment of vitiligo, athlete's foot, ringworm and some chronic skin diseases (*234*), but whether any pharmacological activity must be attributed to α-terthienyl or congeners present in this plant has not been established with certainty.

α-Terthienyl is the only polythiophene for which published information proves biological activity on human skin. Photodermatitis resulting from contact with this compound has been documented (*75*, *234*). The phototoxic reactions were described as biphasic, with burning pain and spreading edema appearing within 10 minutes' exposure to sunlight or UV irradiation. They were followed first by erythema and later by hyperpigmentation persisting for 10 months and localized to the test sites. The relationship between sensitizer dose, ultraviolet light fluence, and skin response is unknown, nor is it known whether individual subjects with different skin types show different responses, or even whether the response is always reproducible in the same or the same type of individuals.

On two occasions, individuals working in the author's laboratory came in contact with accidentally spilled solutions containing either a dithiophene (**6**) or a trithiophene (**T9**). Shortly thereafter they went outdoors and their skin experienced immediate reaction matching the description of the effect reported for the exposure to α-terthienyl. The response due to **6** is documented on a photograph shown below (Fig. 1a). Dr. D. A. Perrine, S. J., demonstrated that skin response could be used as a novel and sensitive detector for polythiophenes. Fractions from the final column chromatographic purification of synthetic **6** had been spotted in duplicate on his forearm. One row was covered with tape and the arm exposed for about 10 minutes to sunlight. The picture, taken a few hours later, unexpectedly revealed the presence of one minor photosensitizing component in addition to **6** (Fig. 1b).

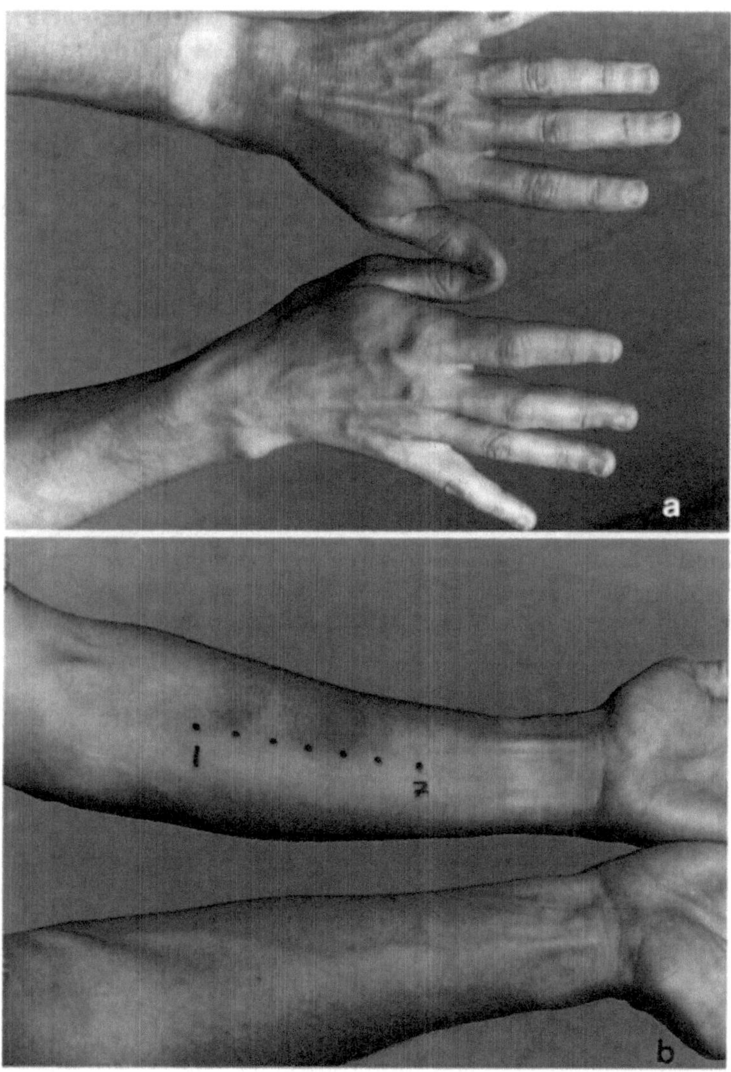

Fig. 1. Skin response to **6**. a), After accidental exposure to the chemical and exposure to sunlight (the outline of the wristwatch worn at the time is clearly visible); b), detection of phototoxic components eluted from column chromatography

In both these cases the skin returned to normal in less than 2 weeks rather than the ten months reported for **T1**. It is quite remarkable that there have been no other published reports of adverse effects on the skin of the scientists involved, despite the large number of studies using bi- and trithiophenes (particularly **T1**).

The cutaneous phototoxicity of α-terthienyl was studied in guinea pig skin, both *in vitro* and *in vivo* (*203*). Effective penetration through the epidermis and superficial dermis produced cutaneous photosensitization comparable to that of intradermally administered α-terthienyl. Phototoxicity was accompanied by a corresponding inhibition of epidermal DNA synthesis in normal and hyperproliferative skin. The authors suggested that α-terthienyl might provide a selective and safer alternative to coal tar and furanocoumarin derivatives for the treatment of psoriasis and other cutaneous diseases by photosensitization. However, no clinical studies in this direction have yet been published.

XI. 4. Hemolysis of Human Erythrocytes

Erythrocytes are a convenient biological test system because a) they are universally available, b) they do not contain any nuclear DNA and c) damage to the integrity of the cell membrane is readily detected by the presence of hemoglobin in solution and by drastic differences in the absorption of 610-nm light by suspensions of intact erythrocytes and by their ghosts. Unfortunately, erythrocytes are not well suited for performing quantitative studies. The response to photosensitizing treatments *in vitro* depends on the age of the erythrocytes: older ones hemolyze more rapidly but give more reproducible results than when freshly drawn (*249*). *In vitro* experiments producing hemolysis of human erythrocytes do not prove, however, that the same chemicals induce similar damage *in vivo*, particularly in the case of damage produced by photosensitization, where light must penetrate tissues before interacting with chemicals in the blood vessels.

The irradiation of erythrocytes in the presence of α-terthienyl leads to hemolysis, and the presence of oxygen is essential (*174*, *185*, *255*, *264*, *267*). However, none of the standard diagnostic tests for the involvement of either singlet oxygen (enhancement in the presence of D_2O and protection by NaN_3) or superoxide ion (protection by superoxide dismutase) were absolutely positive, probably because of the combined effect of the non-homogeneous nature of the reactions and the high hydrophobicity of the chemical. The sensitizer on or in the membrane is effectively out of contact with the hydrophilic reagents used in the tests.

The hemolysis of erythrocytes is but one manifestation of many complex chemical and structural modifications photoinduced by α-terthienyl. The protein components are affected: SDS-acrylamide electrophoresis of irradiated ghosts revealed that faster moving bands were rapidly converted into much slower bands showing no uptake of radioactivity when tritiated α-terthienyl was used (*264*). The lipid components are affected too, as proved by the detection of malondialdehyde, a distinctive product of their oxidative degradation. The integrity of the membrane was gradually modified as indicated by the photoinduced release of potassium ion from the erythrocyte into the medium prior to hemoglobin release. As acetylcholinesterase is a membrane-bound enzyme located on the outside of the erythrocyte membrane, it was not surprisingly inactivated by the photodynamic effect of **T1**, at an initial rate intermediate between K^+ leakage (fastest) and hemoglobin release (*255*). Seven other soluble enzymes were not inactivated significantly, probably because the sensitizer had not crossed the membrane and was therefore not in contact sufficiently close to induce photodynamic damage (*267*).

The membrane of an erythrocyte comprises intricately linked lipid and protein components. Therefore, liposomes made without structural proteins cannot be fully satisfactory model systems. Nevertheless, α-terthienyl induced photodynamic membrane damage in glucose-containing liposomes, probably through lipid peroxidation (*185*). The modification of the membrane properties was revealed by leakage of glucose trapped in the liposomes.

XI. 5. Phototoxicity Towards Bacteria and Fungi

XI. 5. 1. Experimental Results

In their 1947 paper describing the first isolation of a polythiophene from natural sources, that of α-terthienyl from *Tagetes erecta*, Zechmeister and Sease stated that the compound "was void of antibiotic potency against microorganisms like *Staphylococcus aureus*, *Bac. subtilis*, *E. coli* or *Pseudomonas ovalis*" (*270*). One can only regret that their failure to perform the initial microbiological tests in the laboratory near a large window on a sunny day caused a 30-year delay in the discovery that α-terthienyl and other polythiophenes were so highly phototoxic towards so many organisms.

Microorganisms can be grown with great ease, they can be extremely well defined genetically, and many are readily available from reliable sources such as the American Type Culture Collection. The fact that many tolerate anaerobic as well as aerobic conditions makes them invaluable for studying the mechanism of photochemical reactions of organic chemicals *in vivo*. Pure chemicals need not be used, as Daniels first demonstrated with the achene of marigold which inhibited the growth of the yeast *Candida albicans* after irradiation with near-UV light (*84*). Qualitative tests may be performed in Petri dishes containing the proper growth medium immobilized in agar. After thinly spreading a microorganism suspension over the plate, the samples to be tested are placed over the gel. A short incubation period allows diffusion of the active compound(s) into the medium, and subsequent irradiation kills the cells in the vicinity of these active components. During an additional incubation (which can be in the dark) of 12–24 h, the cells unaffected by the treatment undergo further growth and cell division and eventually produce a white surface over the agar, while the areas on the gel where the cells were killed by the photosensitization treatment remain clear. It is convenient to impregnate a small circle of filter paper with a solution of the chemical to be tested, evaporate the solvent, and place the paper over the agar surface. Of course proper controls are required, particularly one in which the procedure is conducted in the absence of ultraviolet light; absence of growth inhibition in the dark coupled with growth inhibition in the presence of light characterizes a phototoxic (*i.e.* photoantibiotic) substance. Growth inhibition in the dark without increased inhibition in the irradiated plate indicates an ordinary antibiotic (in passing, it is important to note that the plastic material of many disposable Petri dishes is not very transparent to ultraviolet light, particularly to short wavelength radiations; it is therefore best to use Pyrex covers over the dishes to be irradiated).

Daniels' method can be combined with chromatography (*9*). When a dried thin-layer chromatography plate is gently placed in contact with the agar surface of a cooled Petri dish diffusion of components occurs from the plate into the agar. The procedure can be repeated using the same plate with several dishes, allowing replica plating. The minimum amount of **T1** detected by this technique was between 1 and 5 µg per spot, using *C. albicans*, *E. coli B/r* and *E. coli* B_{s-1}. Other phototoxic natural products, such as psoralens, could be detected at concentrations up to 100,000 times smaller.

It has been tempting to derive quantitative or qualitative conclusions on relative phototoxicity by measuring the diameters of zones of inhibition produced by different compounds under identical conditions. It is doubtful that these diameters can be used, at the molecular level, to measure directly the toxicity of the compounds tested. They probably reflect the solubility or partition coefficient between lipid and aqueous phases as well as other physical properties of the tested substances at least as much as the quantum efficiencies of the photochemical reactions. This conclusion was recently supported for several compounds (including **T1** and **5**) by a comparison of the relative sizes of inhibition zones with quantitative studies utilizing liquid cultures (*133*). It is noteworthy that one attempt to correlate relative phototoxicity toward *E. coli* and *S. cerevisiae* of several natural products (including **1** and **T1**) in liquid cultures with either water-octanol partition coefficients or photon absorption was not conclusive (*184*).

The initial work in TOWERS' laboratory utilized Daniels' technique (*84*). For example, roots from *Tagetes patula* yielded α-terthienyl (**T1**) and **1**, both of which proved phototoxic to *C. albicans* (*76*). Many Compositae species were tested in the same manner, leaves, stems, roots, and achenes being assayed separately. Most showed no phototoxicity although many produced antibiotic responses. However, some of the plants known to contain bi- and trithiophenes, such as some *Dyssodia* species, did show the now-expected phototoxicity (*238*). In that study, both **1** and **T1** showed phototoxicity to gram-positive (*Bacillus subtilis*, *Staphylococcus albus* and *Streptococcus faecalis*) as well as gram-negative bacteria (*Escherichia coli* and *Proteus vulgaris*). Those results were later confirmed by MARCHANT and COOPER (*178*). Note that *C. albicans* is a well-known pathogen and care must be exercised when handling it. It can be substituted by *C. utilis* (*154*).

In addition to **T1**, **1**, **6** and **7** were also proved phototoxic to *E. coli* as well as *S. cerevisiae* (*98*, *183*, *184*).

Compounds **1** and **T1** were also found to be toxic to a number of phytopathogens and other filamentous fungi (*94*). Complex effects noted with *Alternaria alternata*, *Aspergillus niger*, *Cladosporium variabile*, *Colletotrichum* sp., *Rhizopus nigricans*, *Pythium aphanidermatum*, and *Saprolegnia* sp.. Although some toxicity was noted in dark controls (more with **1**), it was dramatically increased by near-UV irradiation, particularly when the radial growth of somatic mycelia was tested. Timing of the irradiation in relation to treatment with the chemicals was also important. Thus, with conidia of *A. niger* germinating 15–18 h after inoculation onto the growth medium, the radial growth had highest percentage of inhibition of the germ-tubes when the cultures were irradiated 17 h after inoculation. *A. alternata* and *C. variabile* were less sensitive to **T1**, both in the dark and in the light. The latter effect may be due to protection by the dark pigmentation in these organisms.

Marchant and Towers investigated whether fungi isolated from *Bidens* plants containing known photosensitizing components were particularly resistant to them (*179*, *180*). They independently tested the isolated fungi for photosensitivity to **T1** and to several polyacetylenes found in these plants. The growth of the following fungi was found to be inhibited by **T1** in the presence of near-UV light but not in the dark: *Sporobolomyces shibatanus*, *S. salmonicolor*, *Cryptococcus albidus*, *C. laurentii*, *Cladosporium cladosporiodes*, and *Aureobasidium pullulans*. The growth of *R. mucilaginosa* was affected both in the dark and in the light. *S. roseus* and *Epicoccum purpurescens* were not affected by **T1** at all. The results of this study suggested that the organisms did not derive any special protection from contact with their usual photosensitizer-containing hosts. Obviously, effective compartmentalization of the photosensitizers in the plants prevents them from producing their photosensitizing effect *in vivo*.

XI. 5. 2. Mechanistic Studies

In principle, the Daniels' technique should be useful for analyzing the critically important mechanistic point of whether oxygen is required for expressing the photoantibiotic activity of a given sample. However, the difficulty of completely displacing from a gel *all* oxygen molecules by exchange with an inert gas (N_2 or Ar) flowing over the surface should not be underestimated. Because O_2 may be involved in photochemical reactions through a variety of mechanisms (notably physical quenching of excited states, singlet oxygen formation, or superoxide anion formation), the level of exchange sufficient for suppressing the phototoxicity of one chemical is not necessarily low enough for suppressing that of all. Therefore, the lack of phototoxic reaction of one control on a plate does not absolutely prove that truly anaerobic conditions have been reached for all compounds on that plate. The first report on the mechanism of phototoxicity of α-terthienyl derived from this type of experiment with *E. coli* and *C. utilis* found no oxygen dependence (*155*). This publication was mostly based on work submitted by R. Gabriel for his doctoral degree. Unfortunately, it has become absolutely clear that many of the results he claimed could not have been based on actual experiments. This was proved later by Arnason *et al.* with liquid cultures of *E. coli* and *Saccharomyces cerevisiae* (*1*). However, the complete deoxygenation of liquid cultures is itself a difficult task, as demonstrated in the same article which reports that phenylheptatriyne is phototoxic under O_2 as well as under N_2, and more so under the latter. Reinvestigations showed the result to be in error, as this chemical was proved to be phototoxic only in the presence of oxygen (*115*, *160*, *161*).

More detailed mechanistic information was obtained from the inactivation of *E. coli* by α-terthienyl. Gel electrophoresis revealed cross-linking of soluble proteins (*97*). A number of strains were used to probe for possible DNA damage. For example, *rec* mutants of *E. coli* K12 have much reduced ability to repair damaged sites on DNA. Both wild type and mutant strains showed the same size inhibition zones in Daniels' test after photosensitization with α-terthienyl, suggesting that DNA repair is not an important factor in protecting *E. coli* irradiated with this compound (*1*). This point was further investigated in liquid cultures by Tuveson *et al.*, with *E. coli* strains carrying mutations in either the excision repair or the recombination repair mechanisms. No differences were found in the kinetics of inactivation of these strains and no mutations to histidine independence were detected (*239*). All these results support the conclusion that DNA is not an important target for phototoxicity of α-terthienyl *in vivo*.

The conclusion that α-terthienyl and near-UV light principally attack the membrane was further supported by the results obtained with a genetically modified strain of *E. coli* (*240*). Carotenoid pigments, known to quench singlet oxygen very efficiently, can be used as diagnostic compounds for the intermediacy of singlet oxygen in chemical transformations. Genes controlling the synthesis of carotenoid pigments were cloned from *Erwinia herbicola* and expressed in *E. coli*. These cells were completely protected from photosensitization by α-terthienyl (*240*), while control experiments showed no protection against the effect of psoralen derivatives known to have DNA as a target. Since it is believed that the carotenoid pigments produced are associated with the membrane, these experiments strongly support the conclusion that singlet oxygen is responsible for inactivation of *E. coli in vivo*, and that the site(s) of lethal damage are in the membrane. One should note that only partial protection was found with most other photosensitizers tested with the carotenoid producing *E. coli* strain, where a competing superoxide dependent inactivation mechanism was most probably operating.

The relatively small number of microorganisms in which the toxicity of bi- and trithiophenes has been observed forbids extrapolation of the results to other microorganisms with much confidence. Two studies concluded that neither *Pseudomonas aeroginosa* nor *P. fluorescens* were affected by α-terthienyl, either aerobically or anaerobically (*178*, *238*). A third study, however, reported this compound, along with **1**, **6**, and **7**, to be phototoxic to *P. aeroginosa* (*98*). Both **1** and **T1** had the same level of phototoxicity, a level which was greater than for the other two compounds. To add to the confusion, a very recent study again reported that **T1**, **5**, **6**, **24**, and **30** were not phototoxic to *P. aeroginosa* when tested by Daniels' method (*133*). Fortunately, the phototoxicity of **T1** against *Pseudomonas aeroginosa* and *P. fluorescens* has been reinvestigated in liquid cultures in TUVESON's laboratory: these organisms were found to be much more sensitive than *E. coli* in photosensitized treatments with **T1**, with strictly oxygen dependent lethality (*241*).

While the reasons for the variability in the results published in phototoxicity studies with *P. aeroginosa* are unknown, it is interesting to note that the relative phototoxicity of different compounds in different organisms is also variable. HUDSON *et al.* suggested that differences in cell walls or membranes may be responsible and that, for example, yeasts and gram-negative bacteria, having different wall structures, could react differently to thiophene derivatives (*133*). They also noted that eukaryotic cells and viruses show a different spectrum of responses to thiophenes (*132*). Based on the results obtained to date, it is reasonable to expect that many bacteria and fungi will be found to be sensitive to photosensitizing treatments with di- and trithiophenes. The exceptions will probably have very unusual membrane structures and/or accumulate pigments which either act as optical filters or as singlet oxygen quenchers. The latter was already demonstrated in the behavior of one strain of *E. coli* in the presence of **T1** (*240*). The difficulties in predicting phototoxicity in bacteria and fungi are clearly illustrated by the work of DAUB on cercosporin (*85*). This pigment is capable of generating singlet

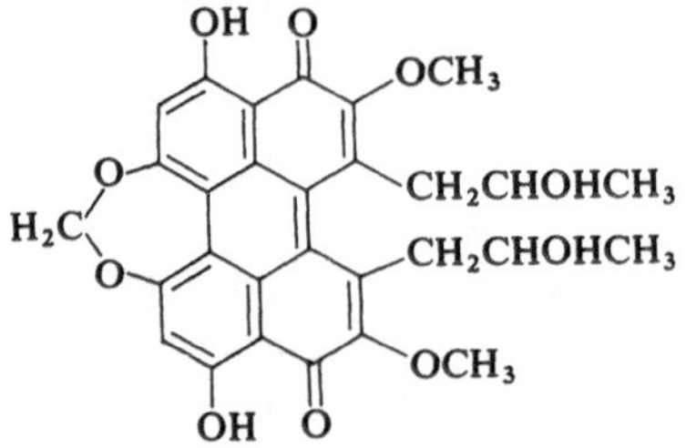

Cercosporin

oxygen as well as superoxide anion photochemically, just like **T1**. It is produced in high concentrations by a fungus which is unaffected by its presence, even in the presence of light. Yeasts, particularly *S. cerevisiae*, are resistant, but *Neurospora crassa* and several *Aspergillus* species are not. It is interesting that several related mycelial fungi which are also plant pathogens (for example *Alternaria*, *Fusarium*, *Colletotrichum*, and

Verticillium species) were not affected by cercosporin. DAUB's work also confirmed that the presence of carotenoids and the composition of the cell wall were important in conferring resistance to organisms, in this case to fungi.

XI. 6. Nematocidal Activity

As related by GOMMERS and BAKKER (*110*) the resistance of *Tagetes* species to root-knot nematodes (*Meloidogyne* species) was observed as early as 1938 (*242*). Experiments in the Netherlands revealed that the cultivation of one crop of marigolds in soil infested with *Pratylenchus penetrans* reduced the nematode population to a level which allowed *Narcissus* bulbs grown in that soil not to suffer from root rot damage normally inflicted by that organism.

UHLENBROEK and BIJLOO observed that, although the juice of *T. patula* was ineffective against cycts of the potato root eelworm, an ethanol extract of the plant had *in vitro* activity against nematodes such as *Ditylenchus dipsaci* and *Anguina tritici*, and larvae of *Heteroda rostochiensis*, *Pratylenchus penetrans*, and *Panagrellus redivivus* (*243*). Further work was done with *T. erecta* roots, from which an active component was isolated and found to be identical with α-terthienyl, the product already found in the petals of this plant (*270*). A second root component, 5-(3-buten-1-ynyl)-2,2′-dithienyl (**1**) was later characterized by the same authors, who found its activity to be lower than that of **T1** against the above nematodes, but high against *Panagrellus redivivus* (*244*). Plants of the family Compositae were later screened extensively for suppressing effect on *P. penetrans*. A large number, 70 out of 150 tested, turned out to have suppressing activity (*109, 111, 114*), including many of the plants belonging to genera known to contain thiophene constituents, e.g. *Berkheya*, *Didelta*, *Eclipta*, *Echinops*, *Flaveria*, *Gaillardia*, and *Tagetes*. These results strongly indicated a correlation between chemical composition and activity against *P. penetrans*. However, several Compositae species which displayed suppressing activity are not known to contain bi- or trithiophenes; perhaps they have not been investigated carefully enough. Unfortunately, marigolds are not active against other kinds of nematodes, some of which even use marigolds as hosts. The same two active compounds isolated from *T. erecta* were also the major components in *T. minuta*, which possesses significant nematocidal activity against *Meloidogyne javanica*, the tobacco root eelworm (*12*, *83*). Root-knot larvae failed to penetrate the roots of marigolds in appreciable numbers, and those that entered did not develop beyond the infective 2 stage larval form. However, **T1** mixed with soil did not control *M. javanica* (*83*). In later work, **T1** (from *T. microglosa* and *T. jalisciencis*) as well as **7** (from *T. microglosa*) were tested against *Meloidogyne incognita*. The former was more active (*188*).

XI. 6. 1. Nematocidal Activity and Mechanistic-Studies in vitro

An accidental discovery in the course of laboratory studies was in large part responsible for the intensive photobiological studies subsequently devoted to bi- and trithiophenes. Populations of nematodes can be kept in Petri dishes, and mechanistic studies on the nematocidal activity of chemicals or plant extracts may therefore be carried out *in vitro*. The suppressing activity of α-terthienyl had been regularly seen in these tests until Gommers inadvertently placed one test Petri dish in a drawer before leaving for the week-end. To his surprise, the nematodes were still alive on his return the next week, when all had been expected to be dead. Rather than dismissing the unanticipated lack of nematocidal activity in this otherwise routine experiment as an abberation, Gommers analyzed the reasons accounting for the observed result and eventually discovered that exposure to light was indispensable for generating the nematocidal activity of α-terthienyl, thus proving it to be a phototoxic molecule.

Further studies clarified some aspects of the light dependence of the activity. The wavelength range of activity was found to be in the near ultraviolet, where α-terthienyl has its λ_{max} at 350 nm, and the extent of activity was clearly related to the length of irradiation. In another piece of pioneering work, Gommers *et al.* proved that α-terthienyl was a singlet oxygen sensitizer, that it could inactivate enzymes, particularly glucose-6-dehydrogenase and acetylcholine esterase, and that the nematode *Aphelenchus avenae*, which can live in the absence of oxygen, was killed by α-terthienyl and ultraviolet light only in the presence of oxygen (*14*, *112*). These results agree with the interpretation that singlet oxygen is also responsible for the death of the nematodes in the *in vitro* sensitized experiments. Later, the light dependence of the nematocidal activity of α-terthienyl was confirmed in adult *Caenorhabditis elegans* (*256*).

XI. 6. 2. Mechanism of the Nematocidal Activity of α-Terthienyl in vivo

The demonstration that α-terthienyl had nematocidal activity *in vitro* and that the activity was light-dependent created an interesting puzzle concerning the mechanism of biological activity *in vivo*. Even though it could be proved that nematodes which had been in contact with the roots of marigolds were killed when exposed to ultraviolet light, the nematodes suppressed by the plants containing α-terthienyl normally remain in the soil and thus probably never see the light of day. Since light transmission through plant roots (*177*) probably can be discounted, it is difficult to argue in favor of a light-dependent mechanism for the toxic effect of α-terthienyl *in vivo*. Yet, no toxicity of α-terthienyl could be demonstrated with any nematodes in the absence of light, so light was clearly essential to the generation of the overall toxic effects.

Gommers and Bakker have recently reviewed the interactions between nematodes and naturally occurring bi- and trithiophenes (*111*) and have presented a model accounting for the nematocidal activity in the dark (*110*). The key element is generation of electronically excited α-terthienyl *in the absence of light* through energy transfer from an electronically excited molecule formed as the result of an enzymatic reaction. The enzymatic reaction utilized to demonstrate the concept was the oxidation of indoleacetic acid with horseradish peroxidase. This reaction involves formation of a dioxetane intermediate which decomposes thermally to produce an electronically excited aldehyde. The latter is quenched by the added α-terthienyl present which thus becoming electronically excited can now react exactly as if it had been generated photochemically. A second energy transfer, this time from α-terthienyl to O_2, then produces singlet oxygen or superoxide anion with their usual toxic reactions. Since it was observed that peroxidase activity in *Tagetes* roots increased markedly in plants infected with *P. penetrans* (*110*), the proposed mechanism is quite relevant to the biological mechanism of toxicity.

Although the location of thiophene derivatives in intact roots has not been determined accurately, sulfur distribution has been analyzed using a micro particle-induced X-ray emission technique (*175*). The lowest concentration was in the epidermis and the highest in the endodermis, providing indirect support to an earlier claim that nematocidal thiophenes were concentrated in the endodermis (*247*). Using a continuous root exudate trapping system and mass spectroscopic analysis, undisturbed roots of *Tagetes patula* were proved to release α-terthienyl and other bithienyl components (*232*). It is interesting that, since some nematodes are insensitive to the effect of α-terthienyl present in plants *in vivo*, the notable lack of selectivity characterizing the phototoxicity of this compound is limited to the reactions which are directly photoinduced.

XI. 6. 3. Nematocidal Activity of Synthetic Analogs

The importance of finding effective nematocides for crop protection amply justifies the synthesis and testing of analogs of compounds with demonstrated activity. Thus, for example, many 2,2'-bithiophenes, trithiophenes, 1,2-dithienylethenes, and analogs where other rings such as benzene, thiazole and pyrazole replace the thiophene nucleus were synthesized and tested on nematodes (*118*, *243*, *245*). The photosensitizing properties of some of these compounds were also investigated *in vitro* (*113*).

XI. 7. Trithiophenes as Insecticides

Although some minor toxic effects of thiophene derivatives can be detected in dark control experiments, the light dependent toxicity of these compounds is their most conspicuous property. The phototoxicity of α-terthienyl and some of its derivatives has been examined in several insects, but practically nothing is known about the effect of bithiophenes. With few exceptions, all the studies to date have dealt with the mere observation of toxic effects, rather than with their mechanism. Most studies were performed under closely controlled laboratory conditions rather than under field conditions.

XI. 7. 1. Effect on Mosquitoes

Mosquitoes are vectors for a number of important diseases, such as malaria, yellow fever and encephalitis. Because these insects readily adapt to environmental pressures and develop resistance to synthetic pesticides, the search for new types of control never ends. The potential for effective use of photosensitizing agents against malaria vectors was recognized by BARBIERI in 1928 (*16*) but little research took place in this area for the next 50 years. Recently, however, several laboratories have investigated the potential of phototoxic agents for insect control. The phototoxicity of α-terthienyl was tested most carefully with various species of *Aedes* mosquitoes, namely *Aedes aegypti* (*7*, *65*, *120*, *151*, *152*, *156*, *159*, *256*), *A. intrudens* (*3*, *201*, *202*), *A. atropalpus* (*2*, *5*, *201*), *A. tritaeniorhynchus* (*65*), *A. epactius* (*120*), as well as with *Culex quinquefasciatus* (*65*, *222*). Although *A. punctor*, *A. decticus*, *A. excrutians*, and *A. flavescens* were also mentioned, their sensitivity to α-terthienyl was not specifically analyzed.

The levels of toxicity have been impressive enough to encourage patenting the use of this chemical (*235*), and to search for possible field control of malaria vectors using crude plant extracts rich in α-terthienyl (*222*) although little is known on phototoxic reactions in *Anopheles* species except for *Anopheles stephensi* (*120*) and *A. gambiae* (*4*). It is not necessary to use a pure chemical or even a plant extract to generate the phototoxicity in mosquito larvae, as crushed root segments of *Tagetes patula* have produced phototoxic response (*256*).

Both artificial UV light and sunlight generate toxicity but, as determined by using filters, the UV portion of the solar spectrum is required for activity (*256*). The action spectrum for α-terthienyl matches almost perfectly the absorption spectrum of this sensitizer, suggesting that the compound itself, rather than a structurally different metabolite, is responsible for the light-dependent toxicity (*7*). Confirmation has not yet been provided (for example by analyzing extracts from treated larvae) for the assumption that unmodified α-terthienyl was accumulated in the organisms.

Before discussing in greater detail the phototoxicity of α-terthienyl in immature mosquitoes, it is important to compare the magnitude of its effect with that of other well-known light-independent insecticides: light-activated α-terthienyl was reported to be less active than Dursban, an organophosphorous compound, but more active than DDT or malathion (*7*). Without light, α-terthienyl was much less toxic.

Little information is available on insect developing resistance to α-terthienyl. Preliminary encouraging studies with *A. aegypti* showed that strains resistant to Dieldrin and to DDT were as sensitive to the phototoxic effects of α-terthienyl as the wild type (*159*). In other experiments, first-instar larvae of *A. aegypti* were treated with α-terthienyl and UV light under conditions where about 50% of the larvae died within 24 h. The surviving organisms were raised, and their eggs collected. First-instar larvae obtained from these eggs were again treated with α-terthienyl and UV light, the process being repeated through 10 generations. No marked differences in the sensitivity of the resulting larvae to photosensitization was detected (*145*). Finally, a recent study found only negligible differences in the phototoxicity of α-terthienyl in *Culex tarsaris* strains which were either susceptible or resistant to malathion (*120*).

Mosquitoes at different stages of development display different sensitivity to photosensitization by α-terthienyl. There is general agreement that adults are not affected, at least to the extent of increasing their mortality. However, little is known about offspring production in treated adult mosquitoes. In order to compare the many reports on the phototoxicity of α-terthienyl to mosquitoes, it is important to appreciate that quantitative data depend on many variables which are difficult to control and to reproduce, either at different times in the same laboratory or from laboratory to laboratory. These variables include the concentration of sensitizer and its mode of application, the type and intensity of the light sources and the timing and duration of the exposures to the chemical and to light. A thorough study determined the survival of eggs, larvae, and pupae of *A. aegypti* (*159*). Hatching was not reduced in eggs irradiated after incubation with α-terthienyl; the irradiation was necessarily of limited duration (30 min) since it had to be stopped before hatching began. Very young larvae were extremely sensitive but less and less sensitive as they become older at the time of treatment. These experiments, using limited incubation with α-terthienyl followed by single exposure to light, revealed that essentially all larvae which survived for 24 h after irradiation developed normally and produced the same percentage of adults as the controls, regardless of the level of initial survival. Pupae are usually very resistant to added chemicals since they no longer feed before turning into adults. However, they were quite susceptible to treatment with α-terthienyl and UV light in the first half of their lifespan; they were no longer affected later. In the case of larvae, at least, the effects recorded in the photosensitized treatments of *A. aegypti* compared qualitatively to those of *A. atropalpus* (*201*).

Experiments with sunlight can never be exactly reproduced, in terms of light intensity and spectral composition as well as other climatic variables such as temperature, relative humidity, etc. The ultimate phototoxicity of α-terthienyl in late 3rd or 4th instar larvae of *A. aegypti*, *A. taeniorhynchus*, and *Culex quiquefasciatus* was very similar under Florida sun (6 h exposure), the LC_{50} values 18 hours later ranging between 0.001 and 0.02 ppm. These values are in reasonable agreement with data obtained under UV irradiation (*120*).

Since most of the studies have been carried out on *Aedes* and *Culex* mosquitoes, it is interesting that α-terthienyl has also been tested as a potential larvicidal agent against a malaria vector with fourth-instar *Anopheles gambiae* in Tanzania (*4*). Little information on the experimental conditions was provided, but the activity of the chemical was found to be somewhat higher than that recorded with *Aedes* mosquitoes in Canada (*3*). These results were believed to compare favorably with those obtained with synthetic pyrethrins.

XI. 7. 2. Absorption, Phototoxic Effects, Elimination and Phototoxicity in Other Insects

Although ovicidal activity could not be demonstrated with α-terthienyl in *A. aegypti* mosquitoes, this effect was discovered in the fruitfly

Drosophila melanogaster (*153*). Eggs placed over a filter paper treated with α-terthienyl failed to hatch after they had been irradiated with near-UV light. The synergistic effect of UV light increased the potency of the chemical almost 10,000-fold. It is tempting to speculate that plants onto which insects had laid eggs might efficiently get protection from being eaten by emerging herbivorous larvae through photosensitized ovicidal response using endogenous bi- or terthienyls. However, this hypothesis has not yet been supported experimentally.

All insect larvae against which the activity of α-terthienyl has been tested were found highly sensitive to photochemical treatment. For example, the phototoxicity of α-terthienyl in blackfly larvae (*256*) and fruitfly larvae (*152*) was recognized early.

The high fluorescence of α-terthienyl facilitates its visual localization in insect tissues. Pictures have shown the fluorescence of the abdominal region, midgut wall, and Malpighian tubules of 4th instar larvae of *Culex tarsaris* treated with α-terthienyl (*120*). Differences in recorded fluorescence intensities in different areas are likely related to differences in the concentration of sensitizer in these tissues but not necessarily proportional to them. However, one does not know the extent to which differential transmission of light (either excitation or fluorescence) through different tissues or the presence of endogenous quenchers must be taken into account.

The level of risk for a given organism in the presence of a phototoxic compound is expected to vary with time. On the one hand, one must be concerned with the rate of input of sensitizer into each cell of the organism where lethal or sub-lethal damage may occur, while on the other hand the organism is likely to respond to the challenge of foreign substances through metabolism and elimination. The quantitative determination of the kinetics of these reactions in rapidly growing organisms is probably hopelessly inextricable, even in the dark, but even more so if photochemical reactions are superimposed.

Tritiated α-terthienyl was used to examine its rate of elimination from *Aedes aegypti* and from *Culex tarsaris* (*120*). Despite the conclusion that an inverse relationship existed between α-terthienyl sensitivity and elimination rate constants, the small differences involved could have resulted from many different factors. Much more useful information will probably be gained by carefully studying the metabolism in a single organism and by analyzing the time dependence and chemical structure of the radioactive molecules released.

The high lipophilicity of α-terthienyl and related molecules probably insures that random distribution in exposed tissues will be achieved by placing an insect larva in an aqueous solution of the sensitizer. Although convenient for qualitative surveys involving the same or very similar organisms, the mere comparison of LC_{50} values for phototoxicity to last-instar larvae in different insects such as *Manduca sexta, Pieris rapae, Heliothis virescens*, and *Ostrinia nubilalis* (*5*) is of dubious mechanistic value. These numbers should be corrected, at the least, to take into account physical differences in weight, size, or surface area.

Topical applications of compounds to selected targets are more apt to provide better focused information (*5, 73, 101, 135, 181*). Topical application of α-terthienyl to larvae of the tobacco hornworm *Manduca sexta* was followed 24 h later by irradiation with UV light. Tissue necrosis occurred at the application sites, and pupae showed morphological abnormalities and irregular melanization (*101*). Larvae fed on a diet containing α-terthienyl were also irradiated. Increased doses of chemicals and irradiation resulted in delayed pupation, reduced pupal weights, increased occurrences of abnormal pupae, and decreased emergence of adult moths. At the highest ingested dose of α-terthienyl (50 μg per g of larval weight, followed by irradiation for 4 h) pupation was delayed about 50%, and there was no subsequent adult emergence.

The effect of dietary α-terthienyl was also tested on larvae of the European corn borer, *Ostrinia nubinalis* (*181*), under conditions using solar light simulating lamps rather than UV lamps. Thus, the UV intensity was probably quite small. The survival to pupation and adult emergence in *O. nubilalis* was significantly reduced. At the highest concentration used (100 μg/g), the mean time to pupation was increased by 30–52% and both the pupation and adult emergence were 71% of the controls. It was also observed that *O. nubilalis* tended to avoid the effect of photosensitization by burrowing into diet or spinning silk (*5*).

Differential insecticidal activity for α-terthienyl in different herbivorous lepidoptera has been noted. Under fluorescent daylight tubes, the topical LD_{50} values for the tobacco hornworm *Manduca sexta*, the tobacco budworm *Heliothis virescens*, and the European corn borer *Ostrinia nubilalis* were in the ratios 1:47:70 (*135*). The toxicokinetics of tritiated α-terthienyl were studied in these three insects. Following either oral or topical administration, larvae of *H. virescens* and *O. nubilalis* rapidly cleared the chemical from the body via the feces, but *M. sexta* did more slowly. For example, after 48 h of feeding on treated diet, the ratio of radiolabel in the body to that in feces was 16:84 for *O. nubilalis*, 32:68 for *H. virescens*, and 58:42 for *M. sexta*. Likewise, the half-time values for elimination of ^3H after topical application in *O. nubilalis, H. virescens*, and *M. sexta* were approximately in the ratios 1:3:6. Unfortunately, only total radioactivity was measured and the chemical structure(s) of the radioactive species excreted was not disclosed.

XI. 7. 3. Mechanism of Photoinsecticidal Activity

Although it has been assumed that the photoinsecticidal activity of α-terthienyl is related to the molecule's ability to generate singlet oxygen, no direct experimental support has so far been obtained. One major obstacle is the difficulty of proving that oxygen itself is involved in the toxic reactions as the insects cannot survive anaerobic control experiments, even in the dark.

At the molecular level, little is known about the target(s) of the lethal photosensitized reactions of α-terthienyl in insects. The only report to date concerns the inactivation of the important enzyme acetylcholinesterease in larvae of the mosquito *A. aegypti* (*156*). After confirming that, *in vitro*, pure acetylcholinesterase rapidly decreased in activity upon irradiation (an oxygen-dependent process), the total soluble enzyme activity was tested in larvae irradiated for various lengths of time after initial incubation in the presence of α-terthienyl. A gradual decrease in enzyme activity was observed; about 40% of the initial enzyme activity was lost at the point where all the larvae had died. Many other enzymes in the larvae are probably inactivated as a result of photosensitized treatments and the decrease of acetylcholinesterase activity may not be the sole direct cause of the death of the larvae.

XI. 7. 4. Feeding Deterrence

α-Terthienyl added to an insect diet is not necessarily taken up by the insect for which it is intended. Instead, its presence may deter the insect from feeding on the prepared diet, as was demonstrated with the larvae of the dark-sided cutworm *Euxoa messoria* (*73*). The efficiency of diet conversion to insect biomass was markedly decreased when α-terthienyl was added, and UV did not enhance this effect. The combined effect of UV light and the chemical added to the diet and was much greater than either one alone. Although initially the larvae ate very little, they eventually started to feed normally on the artificial diet. Their growth was slow and all died before molting to the next instar. As these experiments were all performed with artificial diets rather than with plants tissues known to contain α-terthienyl, the assertion that α-terthienyl has an important function in protecting plants that contain it from insect herbivores (*73*, *256*) still remains unverified experimentally, as already mentioned above for the ovicidal activity.

XI. 7. 5. Is α-Terthienyl Suitable as a Practical Insecticide?

The logical sequence in studying the insecticidal activity of an active compound should be to proceed from small scale laboratory experiments to increasingly larger field tests. There, not only must the activity against a selected pest be sufficient to justify the cost of manufacture, but the compound must also be proved to be environmentally acceptable. The criteria may depend on current governmental legislation, but the ideal goal should be to secure compounds with a very narrowly focused range of activities against natural organisms.

Field trials of α-terthienyl compared the results in natural and artificial ponds (*202*). In addition to establishing concentration levels and formulation methods for controlling the target mosquito *A. intrudens,* the effect on non-target organisms was assessed. The survival of midge (Diptera, Chaoboridae), caddisfly (Trichoptera, Limnephidae), and damselfly (Odonata, Libellulidae) larvae, as well as freshwater shrimp (*Chirocephalopsis bundyi*) was evaluated in the field. Only the midge larvae were thought likely to be affected at the application rates required to control mosquito larvae. Daphnia, snails, and rainbow trout (*Salmo gairdneri*) were claimed to tolerate a photosensitized α-terthienyl treatment under laboratory conditions.

The suggestion that the insecticide activity of α-terthienyl would not harm important non-target organisms conflicts with reports that this compound *in vitro* is highly phototoxic in Daphnia (*151*), fish (*Pimephales promelas*, fathead minnow) (*158*) and late embryonic stages of *Rana pipiens* (*157*) and *Hyla crucifer* (*145*). Actually, the phototoxicity in this fish makes α-terthienyl one of the most potent fish poisons known.

It is clear that α-terthienyl has exhibited phototoxicity in all organisms in which it has been tested to date. This lack of specificity suggests that the prospects for safe and useful applications of this substance as an insecticide will remain poor until specifically targeted modes of delivery are devised or suitable structural modifications are achieved. Although a number of substituted α-terthienyl molecules have been evaluated for phototoxicity (*4*), no analyses of the relationship between structure, activity, and selectivity have so far provided any meaningful guides to the design of vastly superior insecticides.

Finally, the persistence of α-terthienyl in the environment is not known accurately. The mere analysis of the change with time of the spectrum of an aqueous solution to which a known amount of α-terthienyl was added is very deceptive because of the very limited solubility of the chemical. Aggregation occurs when the solubility limits are exceeded, and the spectral data (absorption or fluorescence) clearly show the rapid change (*150*). However, the decrease in spectral intensity cannot be equated with disappearance of the substance as a chemical entity. The presence of essentially all the original material can be proved by comparing the spectrum of the solution obtained by addition of a good solvent in sufficient quantity, such as ethanol, to that of the corresponding control.

XI. 8. Toxicity Toward Fish and Other Aquatic Organisms

XI. 8. 1. Toxicity Toward Fish

α-Terthienyl is a powerful fish poison. Its toxicity was tested in the laboratory with fathead minnows (*Pimephales promelas*) about 5 cm long (*158*). No adverse effects occurred to fish kept in the dark in water containing α-terthienyl up to 4×10^{-6} *M*. In the presence of either sunlight or near-UV light, however, the fish died very rapidly. Using identical fish, α-terthienyl with only 30 min of exposure to light ($LC_{50} = 0.7 \times 10^{-7}$ *M*) was essentially as toxic as rotenone, probably the best-known fish poison ($LC_{50} = 1 \times 10^{-7}$ *M*), and endrin, a polychlorinated hydrocarbon ($LC_{50} = 3 \times 10^{-7}$ *M*).

The conditions for maximum phototoxicity of α-terthienyl have not been established. It is likely that the toxic effect will be even greater in fish, particularly younger ones, exposed to light for longer durations. Thus the long term survival of natural fish populations exposed to the chemical might be severely affected. How the fish come in contact with the chemical is also important. For example, fish eating Daphnia previously treated with α-terthienyl had a much higher survival rate after irradiation than fish directly exposed to the chemical as in the above experiments (*158*).

A much lower sensitivity has been reported in tests with other fish (*202*). Probably the size of the fish as well as their species will affect their sensitivity to the photosensitizer, but no detailed studies in this area have been reported.

Another point of interest in relation to possible applications of α-terthienyl as a biocide is the relationship between initial exposure of the organism to the chemical and its irradiation. In laboratory experiments, fish exposed to α-terthienyl at a concentration leading to 50% survival when irradiated 30 min later survived better when they were transferred and kept in pure water before irradiation. The survival was total after a stay of only 3 h in clean water. It is not known whether metabolism or other modes of depuration (i.e. loss of active compound through undetermined means) are responsible for the decreasing toxicity of the sensitizer.

The duration of the fish incubation with the chemical prior to irradiation also affects the phototoxicity of α-terthienyl. In the experimental conditions selected, the survival was 100% when the fish were irradiated for 30 min immediately upon contact with the solution, 0% when irradiated after 2 h, and 70% when the irradiation was started 8 h after initial contact. This indicated that a) a minimum incubation period was required for the sensitizer to reach the biological targets leading to lethal damage, and b) depuration took place. However, nothing is known about the relationship between the initial concentration of sensitizer and the magnitude of this phenomenon (*151*).

As mentioned in Section XI.7.5, spectroscopic monitoring revealed rapid modification of the physical properties of α-terthienyl dispersed in water (*150*). Preliminary experiments tested whether these modifications were associated with differences in photobiological properties. The test consisted in adding a small amount of sensitizer (in ethanol) to produce a 5×10^{-7} *M* concentration and immediately allowing fish to swim for 1 h prior to 1 h of irradiation with near-UV light. Twenty four hours later, 95% of the fish were dead. Fish introduced into ageing solutions survived the phototreatment increasingly better, to the point where all survived when treated in a 48-h old dispersion, even though it could be shown that all the α-terthienyl was still present in such a dispersion (*145*). This observation that the phototoxicity of α-terthienyl is critically dependent upon the state of dispersion of the sensitizer may explain differences in phototoxicity noted for the same organisms in different laboratories. It should also provide an important component to the design of strategies for pest control with α-terthienyl.

XI. 8. 2. Phototoxicity Toward Tadpoles

Late embryonic stages of *Rana pipiens* (*157*) and *Hyla crucifer* (*151*) were found to be extremely sensitive to the effect of α-terthienyl *in vitro*, but only in the presence of sunlight. The phototoxicity level was comparable to that recorded with first-instar larvae of the mosquito *Aedes aegypti* (*159*). The relationship between developmental stage and sensitivity to photosensitized damage by α-terthienyl has not been investigated.

XI. 8. 3. Phototoxicity Toward Daphnia

α-Terthienyl was highly phototoxic toward Daphnia in the laboratory (*19*). Here again, the effect of the various parameters on the level of phototoxicity has not been systematically investigated. Exogenous β-carotene, an efficient singlet oxygen quencher, gave only limited protection in the photosensitized treatment, probably acting more as a light screening agent than as a quencher of excited states. Actually, at the concentration levels of most noticeable protection, β-carotene turned out to be toxic to the Daphnia in the dark.

α-Terthienyl is probably toxic to many other aquatic organisms in the presence of UV light, as judged by preliminary laboratory experiments with snails and other unidentified specimens from a local pond, but much additional work in this area would be needed for satisfactory quantitative results.

XI. 8. 4. Phototoxicity Toward Cercariae

A number of aquatic organisms are dangerous human parasites. The cercariae of the digenetic trematodes which are responsible for cercarial dermatitis (swimmers' itch) were considered attractive targets for application of the phototoxic properties of α-terthienyl (*116*, *236*). The studies evaluated whether α-terthienyl damaged free-swimming cercariae. Both a leptocercous echinostomatid and a furcocercous schistosomatid obtained from freshwater snails were tested. Although an effect of the chemical in the dark was noted, irradiation with UV light greatly decreased the time required for immobilization of the cercariae. Graham *et al.* also attempted in these studies to sensitize the cercariae by keeping powdered roots of *Tagetes patula* in the water containing snails (*Physa occidentalis*) producing cercariae. After 72 h, all the snails survived and released active cercariae in abundance (interestingly, 50% of the snails kept in pure water as controls had died). Irradiation with UV light alone did not affect either the snails or the cercariae, whereas cercariae introduced into a 72-h infusion of roots were rapidly killed upon irradiation. Additional experiments designed to analyze the relationship between phototoxic responses and incubation time, as described above with fish, would help to clarify the results obtained with the cercariae.

Although an infusion of *Tagetes* roots helps maintaining snails for cercarial production in the laboratory, there are no experimental data implicating thiophene components at all in this phenomenon.

XI. 9. Toxicity of α-Terthienyl Toward Plants

In a patent filed in 1959 α-terthienyl was claimed to help control plant growth (*119*). The patent asserted that the compound could be directly applied to the plants whose growth was to be controlled or that the application could be made in advance of an anticipated weed infection. Formulated as a wettable powder composition and applied at a rate of 8 lb/acre, an excellent control of annual broadleaf weeds and grasses was claimed. A dose five times higher provided excellent control of deep-rooted perennial weeds such as bindweed and evening primrose. The mechanism of this herbicidal action is not yet clearly understood, although it is tempting to suggest that light-dependent effects are involved.

The effect of α-terthienyl on plant growth was later studied more specifically in the laboratory with two common weeds, *Asclepias syriaca* and *Chenopodium album,* and two common forage species, *Phleum pratense* and *Trifolium pratense* (*6*). Exposure to near-UV was required for demonstrating activity. The studies, carried out *in vitro* with seedlings, revealed that the first named species was the most sensitive. Seeds treated with a solution of α-terthienyl also suffered reduced germination *in vitro*. The potential allelopathic character of *Tagetes* was demonstrated by characterizing α-terthienyl in extracts of soil taken from pure stands of these plants. However, the allelopathic activity of plants containing α-terthienyl or other thiophene-containing photosensitizers is yet to be demonstrated in the field. Many synthetic analogs and derivatives of α-terthienyl have been tested for herbicidal activity, but none appear to be competitive with commercial products currently on the market.

α-Terthienyl was also shown to inhibit growth of the alga *Chlorella pyrenoidosa,* but only in the presence of ultraviolet light (*6*). Although the same study reported that a number of phototoxic acetylenic molecules affected the growth of several marine and other freshwater algae in the presence of UV light, α-terthienyl was not tested with these species. The only other report indicated that oxygen evolution in the light was reduced in *Chlorella vulgaris* which had been preirradiated in the presence of α-terthienyl, but oxygen uptake in the dark was not affected as much (*219*).

XII. Conclusion

A much greater diversity of chemical structures has been reported for bithiophenes than for trithiophenes. Isolation and structure determination of most of these substances can be credited to BOHLMANN'S laboratory. In contrast, surprisingly little is known about the biological activity of bithiophenes, as most of the investigations have been performed with α-terthienyl. This compound displays an impressive array of cytotoxic properties which are practically all light-dependent. The very high level of toxicity elicited by α-terthienyl toward individual targets raised high hopes that practical applications could be devised for the specific control of such diverse organisms as weeds, mosquito larvae, sand flies, cercariae, fish, etc. However, the cumulative impact of this diversity of biological targets makes the use of α-terthienyl unlikely for large scale applications. The environmental risks appear overwhelmingly high unless modes of delivery specifically targeted toward one species become available. The reasons for the unusually high phototoxicity of α-terthienyl are not fully known. Its ability to generate singlet oxygen efficiently is certainly one reason, but it cannot be the only one, as other substances capable of generating singlet oxygen almost quantitatively are less biologically active. Perhaps the well-known chemical resistance of thiophenes toward reaction with singlet oxygen protects α-terthienyl from self-destruction to a greater extent than other sensitizers. Other factors such as solubility or resistance to metabolic transformations may also have to be considered.

It would be interesting to know whether other naturally-occurring bithiophenes have activities which rival those of α-terthienyl. However, even if no bi- or trithiophenes ever gain practical applications, α-terthienyl will have been a most important reference compound for understanding and defining those photophysical and photochemical reactions which are responsible for the phenomenon of phototoxicity.

Acknowledgement

The financial support of the Public Health Service and the University of Illinois at Chicago greatly facilitated the studies on phototoxicity in the author's laboratory. A Visiting Professorship at the Museum National d'Histoire Naturelle in Paris, and a Fulbright Fellowship are also gratefully acknowledged. The author is also indebted to the many colleagues who generously shared their knowledge of various fields of biology and physical chemistry, and to the undergraduate, graduate, and postdoctoral associates who participated in the experimental work in his laboratory.

References

1. ARNASON, T., G.F.Q. CHAN, C.K. WAT, K. DOWNUM, and G.H.N. TOWERS: Oxygen Requirement for Near-UV Mediated Cytotoxicity of α-Terthienyl to *Escherichia coli* and *Saccharomyces cerevisiae*. Photochem. Photobiol. **33**, 821 (1981).
2. ARNASON, J.T., B.J.R. PHILOGENE, C. BERG, A. MACEACHERN, J. KAMINSKI, L.C. LEITCH, P. MORAND, and J. LAM: Phototoxicity of Naturally Occurring and Synthetic Thiophene and Acetylene Analogues to Mosquito Larvae. Phytochem. **25**, 1609 (1986).
3. ARNASON, J.T., B.J.R. PHILOGENE, F. DUVAL, C.W. BERG, S. IYENGAR, and P. MORAND: Efficacy of Formulations of the Phototoxic Insecticide, α-Terthienyl Towards *Aedes* spp. Bioact. Mol. (Chem. Biol. Nat.-Occurring Acetylenes Relat. Compd.) **7**, 305 (1988).
4. ARNASON, J.T., B.J.R. PHILOGENE, P. MORAND, K. IMRIE, S. IYENGAR, F. DUVAL, C. SOUCY-BREAU, J.C. SCAIANO, N.H. WERSTIUK, B. HASSPIELER, and A.E.R. DOWNE: Naturally Occurring and Synthetic Thiophenes as Photoactivated Insecticides. ACS Symp. Ser. **339**, 164 (1989).
5. ARNASON, J.T., B.J.R. PHILOGENE, P. MORAND, J.C. SCAIANO, N. WERSTIUK, and J. LAM: Thiophenes and Acetylenes: Phototoxic agents to Herbivorous and Blood-feeding Insects. ACS Symp. Ser. **339**, 255 (1987).
6. ARNASON, T., J.R. STEIN, E. GRAHAM, C.K. WAT, G.H.N. TOWERS, and J. LAM: Phototoxicity to Selected Marine and Freshwater Algae of Polyacetylenes from Species in the Asteraceae. Can. J. Bot. **59**, 54 (1981).
7. ARNASON, T., T. SWAIN, C.-K. WAT, E.A. GRAHAM, S. PARTINGTON, G.H.N. TOWERS, and J. LAM: Mosquito Larvicidal Activity of Polyacetylenes from Species in the Asteraceae. Biochim. Syst. Ecol. **9**, 63 (1981).
8. ASANO, T., S. ITO, N. SAITO, and K. HATAKEDA: A Simple Synthesis of 2,2′,5′,2″-Terthienyl. Heterocycles **6**, 317 (1977).
9. ASHWOOD-SMITH, M.J., G.A. POULTON, O. CESKA, M. LIU, and E. FURNISS: An ultrasensitive Bioassay for the Detection of Furanocoumarins and Other Photosensitizing Molecules. Photochem. Photobiol. **38**, 113 (1983).
10. ATKINSON, R.E.: Bi- and Terthienyl Compounds as Fluorescent Whiteners. Brit. GB 1273986 10 May 1972.
11. ATKINSON, R.E., R.F. CURTIS, and G.T. PHILLIPS: Bithienyl Derivatives from *Tagetes minuta* L. Tetrahedron Lett. **1964**, 3159.
12. — — —: Naturally-Occurring Thiophenes. Bithienyls from *Tagetes minuta* L. J. Chem. Soc. **1965**, 7109.
13. — — —: Naturally Occurring Thiophenes. IV. Synthesis of some 2,2′-Bithienyl Derivatives from Cuprous Acetylides. J. Chem. Soc. (C), **1967**, 2011.
14. BAKKER, J., F.J. GOMMERS, I. NIEUWENHUIS, and H. WYNBERG: Photoactivation of the Nematicidal Compound α-Terthienyl from Roots of Marigolds (*Tagetes* Species). A Possible Singlet Oxygen Role. J. Biol. Chem. **254**, 1841 (1979).
15. BALDWIN, J.E. Rules for Ring Closure. J.C.S. Chem. Commun. **1976**, 734.
16. BARBIERI, A.: Sensibilizadores Fluorescentes como Larvicidas. Accion Fotodinamica de la Luz. Riv. Malariol. **7**, 456 (1928).
17. BASTOS, M.M.S.M., A. KIJOA, J.M. CARDOSO, A.B. GUTIERREZ, and W. HERZ: Lignans and Other Constituents of *Centaurea sphaerocephala* ssp. *polyacantha*. Planta Med. in press (1990).
18. BAZIN, M., R. SANTUS, and J. KAGAN: Unpublished results.
19. BENNETT, W.J., J.L. MAAS, S.A. SWEENEY, and J. KAGAN: Phototoxicity in Aquatic Organisms: the Protecting effect of Beta-Carotene. Chemosphere **15**, 781 (1986).

20. BESTMANN, H.J., and W. SCHAPER: Reaktionen von Thioacylalkylidentriphenylphosphoranen-Eine neue Thiophensynthese. Tetrahedron Lett. **1979**, 243.
21. BIRNBAUM, D., and B.E. KOHLER: Lowest Energy Excited Singlet State of 2,2:5′,2″-Terthiophene, an Oligomer of Polythiophene. J. Chem. Phys. **90**, 3506 (1989).
22. BOHLMANN, F., M. AHMED, R.M. KING, and H. ROBINSON: Polyacetylenic Compounds. Part 262. Acetylenic Compounds from *Bidens graveolens*. Phytochemistry **22**, 1281 (1983).
23. BOHLMANN, F., W.R. ABRAHAM, R.M. KING, and H. ROBINSON: Polyacetylenic compounds. Part 257. Thiophene Acetylenes and Flavanols from *Pterocaulon virgatum*. Phytochemistry **20**, 825 (1981).
24. BOHLMANN, F., C. ARNDT, K.M. KLEINE, and H. BORNOWSKI: Polyacetylenic Compounds. 69. The Acetylene Compounds of the Genus *Echinops*. Chem. Ber. **98**, 155 (1965).
25. BOHLMANN, F., C. ARNDT, K.M. KLEINE, and M. WOTSCHOKOWSKY: Polyacetylenic Compounds. 75. New Constituents from *Bidens* Species. Chem. Ber. **98**, 1228 (1965).
26. BOHLMANN, F., R.N. BARUAH, and X. DOMINGUEZ: A Further Dithienyl Derivative from *Porophyllum scoparia*. Planta Med. **1**, 77 (1985).
27. BOHLMANN, F., and E. BERGER: Polyacetylenic Compounds. 73. The Polyynes of the Genus *Buphthalmum*. Chem. Ber. **98**, 883 (1965).
28. BOHLMANN, F., and E. BRESINSKY, Polyacetylenic Compounds. 120. Conversion of Reactive Acetylene Compounds with Sulfur Compounds. Chem. Ber. **100**, 107 (1967).
29. BOHLMANN, F., T. BURKHARDT, and C. ZDERO: Naturally Occurring Acetylenes, New York: Academic Press (1973).
30. BOHLMANN, F., et al.: Unpublished results in reference (*29*).
31. BOHLMANN, F., et al.: Unpublished results in reference (*57*).
32. BOHLMANN, F., U. FRITZ, R.M. KING, and H. ROBINSON: Naturally Occurring Terpene Derivatives. Part 301. Fourteen Heliangolides from *Calea* Species. Phytochemistry **20**, 743 (1981).
33. BOHLMANN, F., and M. GRENZ: Naturally Occurring Terpene Derivatives. Part 172. A New Germacranolide from *Munnozia maronii*. Phytochemistry **18**, 334 (1979).
34. BOHLMANN, F., M. GRENZ, M. WOTSCHOKOWSKY, and E. BERGER: Polyacetylene Compounds. 134. New Thiophenacetylene Compounds. Chem. Ber. **100**, 2518 (1967).
35. BOHLMANN, F., R.K. GUPTA, R.M. KING, and H. ROBINSON: Naturally Occurring Terpene Derivatives. Part 448. Two Furanoheliangolides from *Calea angusta*. Phytochemistry **21**, 2117 (1982).
36. BOHLMANN, F., and P. HERBST: Polyacetylenic Derivatives. 43. The Constituents of *Tagetes* Species. Chem. Ber. **95**, 2945 (1962).
37. BOHLMANN, F., and U. HINZ: Polyacetylenic Compounds. 72. Biogenetic Conversions of Tridecenepentayne. Chem. Ber. **98**, 876 (1965).
38. BOHLMANN, F., J. JAKUPOVIC, H. ROBINSON, and R.M. KING: Polyacetylene Compounds. Part 258. A Dithienylacetylene from *Porophyllum ruderale*. Phytochemistry **19**, 2760 (1980).
39. BOHLMANN, F., and K.M. KLEINE: Polyacetylenic Compounds. 47. The Polyynes from *Flaveria rependa*. Chem. Ber. **96**, 1229 (1963).
40. BOHLMANN, F., K.M. KLEINE, and C. ARNDT: Polyacetylene Compounds. 57. Naturally Occurring Thiopheneacetylene Compounds. Chem. Ber. **97**, 2125 (1964).
41. BOHLMANN, F., N. LE VAN, and J. PICKARDT: Naturally Occurring Terpene Derivatives, 108. An Anomalous Sesquiterpene from *Berkheya radula* (Harv.) De Willd. Chem. Ber. **110**, 3777 (1977).
42. BOHLMANN, F., M. LONITZ, and K.H. KNOLL: New lignan derivatives from the Heliantheae family. Phytochemistry **17**, 330 (1978).

43. BOHLMANN, F., and D. MOHAMMADI: A Further Bithienyl Derivative from *Berkheya zeyheri*. Phytochemistry **22**, 2856 (1983).
44. BOHLMANN, F., D. MOHAMMADI, and J. JAKUPOVIC: Sesquiterpene Lactones from *Berkheya* species. Planta Med. **50**, 192 (1984).
45. BOHLMANN, N. LE VAN, V.C.P. THI, J. JACUPOVIC, A. SCHUSTER, V. ZABEL, and W.H. WATSON: Naturally Occurring Terpene Derivatives. 226. β-Isocomene, a New Sesquiterpene from *Berkheya* species. Phytochemistry **18**, 1831 (1979).
46. BOHLMANN, F., K.M. RODE, and C. ZDERO: Polyacetylenic Compounds. 121. Polyyines from Helenieae Tribe. Chem. Ber. **100**, 537 (1967).
47. BOHLMANN, F., and A. SUWITA: Polyacetylenic compounds. 231. Further Constituents from Species of the Tribe Arctotideae. Chem. Ber. **108**, 515 (1975).
48. ——: A new Guaianolide and a Secoguaianolide from *Helicrysum splendidum*. Phytochem. **18**, 885 (1979).
49. BOHLMANN, F., M. WOTSCHOKOWSKY, U. HINZ, and W. LUCAS: Polyacetylenic Compounds. 95. Biogenesis of Certain Thiophene Compounds. Chem. Ber. **99**, 984 (1966).
50. BOHLMANN, F., and C. ZDERO: Polyacetylene Compounds. 173. Constituents of *Eclipta erecta*. Chem. Ber. **103**, 834 (1970).
51. ——: Polyacetylene Compounds. 190. Constituents of *Buphthalmum salicifolium*. Chem. Ber. **104**, 958 (1971).
52. ——: Polyacetylene Compounds. 205. On the Constituents of the Tribe Arctotideae. Chem. Ber. **105**, 1245 (1972).
53. ——: Polyacetylenic Compounds, 238. On the Constituents of the genus *Dyssodia*. Chem. Ber. **109**, 901 (1976).
54. ——: On the Constituents of the Tribe Mutisieae. Phytochem. **16**, 239 (1977).
55. ——: New Germacrolides from *Platycarpha glomerata*. Phytochem. **16**, 1832 (1977).
56. ——: Naturally Occurring Terpene Derivatives. Part 173. Dimeric Terpene Ketones from *Tagetes gracilis*. Phytochemistry **18**, 341 (1979).
57. ——: Naturally Occurring Thiophenes. In: The Chemistry of Heterocyclic Compounds. Vol 44: Thiophenes and its Derivatives (Gronowitz, S., ed.), p 261. New York: John Wiley & Sons (1985).
58. BOHLMANN, F., C. ZDERO, W.R. ABRAHAM, A. SUWITA, and M. GRENZ: Naturally Occurring Terpene Derivatives. Part 256. New Diterpene and new Dihydrochalcone Derivatives Together with Further Components from *Helichrysum* species. Phytochemistry **19**, 873 (1980).
59. BOHLMANN, F., C. ZDERO, and W. GORDON: Polyacetylene Compounds. 123. Components of *Berkheya adlami*. Chem. Ber. **100**, 1193 (1967).
60. BOHLMANN, F., C. ZDERO, and M. GRENZ: Polyacetylene Compounds. Part 241. Constituents of some Genera of the Tribes Helenieae and Senecioneae. Phytochemistry **15**, 1309 (1976).
61. BOHLMANN, F., C. ZDERO, R.M. KING, and H. ROBINSON: Naturally Occurring Terpene Derivatives. Part 469. Thymol Derivatives from *Porophyllum riedelii*. Phytochemistry **22**, 1035 (1983).
62. BOHLMANN, F., C. ZDERO, and N. LE VAN: Naturally Occurring Terpene Derivatives. Part 168. New Geranylcoumarin Derivatives and Further Constituents of the tribe Mutisieae. Phytochemistry **18**, 99 (1979).
63. BOHLMANN, F., C. ZDERO, and P. MAHANTA: Naturally-Occurring Terpene Derivatives. Part 105. New Diterpenes from *Dimorphotheca* and *Viguiera* Species. Phytochemistry **16**, 1073 (1977).
64. BOHLMANN, F., C. ZDERO, and M. SILVA: Two Further Eremophylane Derivatives from *Tessaria absynthioides*. Phytochem. **16**, 1302 (1977).

65. BOROVSKY, D., J.R. LINLEY, and J. KAGAN: Polycyclic Aromatic Compounds as Phototoxic Larvicides. J. Am. Mosq. Control Assoc. **3**, 246 (1987).
66. BRYZGIS, M., S.N. DHAWAN, J. KAGAN, K. REID, S.P. SINGH, and L. TOW: The Reaction of Phosphorus Pentachloride with 2-Acetylthiophene and Acetophenone. J. Org. Chem. **48**, 703 (1983).
67. CAMPBELL, G., J.D.H. LAMBERT, T. ARNASON, and G.H.N. TOWERS: Allelopathic Properties of α-Terthienyl and Phenylheptatriyne. Naturally Occurring Compounds from Species of Asteraceae. J. Chem. Ecol. **8**, 961 (1982).
68. CARPITA, A., and R. ROSSI: Palladium-Catalyzed Syntheses of Bi- and Terthiophenes. Gazz. Chim. Ital **115**, 575 (1985).
69. CARPITA, A., R. ROSSI, and C.A. VERACINI: Synthesis and Carbon-13 NMR Characterization of Some π-Excessive Heteropolyaromatic Compounds. Tetrahedron **41**, 1919 (1985).
70. CASTRO, A. V., and C.O. CASTRO: Natural Derivatives of Thiophene in the Root of *Tagetes microglosa*. Rev. Latinoam. Quim. **9**, 204 (1978).
71. CASTRO, C.O., and C.L. MUNOZ: Natural Thiophene Derivatives From the Roots of *Tagetes jalisciencis*. Rev. Latinoam. Quim. **13**, 36 (1982).
72. CHALLENGER, F., and J.L. HOLMES: The Orientation of Substitution in the Isomeric Thiophthens. The Synthesis of Solid Thiophthen [Thiopheno(3′:2′-2:3)thiophen]. J. Chem. Soc. **1953**, 1837.
73. CHAMPAGNE, D.E., J.T. ARNASON, B.J.R. PHILOGENE, G. CAMPBELL, and D.G. MCLACHLAN: Photosensitization and Feeding Deterrence of *Euxoa messoria* (Lepidoptera: Noctuiidae) by α-Terthienyl, a Naturally Occurring Thiophene from the Asteraceae. Experientia **40**, 577 (1984).
74. CHAN, G.F.Q., M.M. LEE, J. GLUSHKA, and G.H.N. TOWERS: Photosensitizing Thiophenes in *Porophyllum*, *Tessaria* and *Tagetes*. Phytochemistry **18**, 1566 (1979).
75. CHAN, G.F.Q., M. PRIHODA, G.H.N. TOWERS, and J.C. MITCHELL: Phototoxicity Evoked by Alpha-Terthienyl. Contact Dermatitis **3**, 215 (1977).
76. CHAN, G.F.Q., G.H.N. TOWERS, and J.C. MITCHELL: Ultraviolet-Mediated Antibiotic Activity of Thiophene Compounds of *Tagetes*. Phytochemistry **14**, 2295 (1975).
77. CHODKIEWICZ, W.: Contribution to the Synthesis of Acetylenic Compounds. Ann. Chimie **1957**, 819.
78. COOPER, G.K., and C.I. NITSCHE: Alpha-Terthienyl, Phototoxic Allelochemical. Review of Research on its Mechanism of Action. Bioorg. Chem. **13**, 362 (1985).
79. COREY, E.J., and P.L. FUCHS: A Synthetic Method for Formyl → Ethynyl conversion. Tetrahedron Lett. **1972**, 3769.
80. COSIO, E.G., R.A. NORTON, E. TOWERS, A.J. FINLAYSON, E. RODRIGUEZ, and G.H.N. TOWERS: Production of Antibiotic Thiarubrines by a Crown Gall Tumor Line of *Chaenactis douglasii*. J. Plant Physiol. **124**, 155 (1986).
81. COUTURE, A., and A. LABLACHE-COMBIER: Thiophene Chemistry. Chem. Commun. **1969**, 524.
82. CROES, A.F., M. BOSVELD, and G.J. WULLEMS: Control of Thiophene Accumulation in *Tagetes*. Bioact. Mol. (Chem. Biol. Nat.-Occurring Acetylenes Relat. Compd.) **7**, 255 (1988).
84. DANIELS, Jr., F.: A Simple Microbiological Method for Demonstrating Phototoxic Compounds. J. Invest. Dermatol. **44**, 259 (1965).
85. DAUB, M.E.: The Fungal Photosensitizer Cercosporin and its Role in Plant Disease. A.C.S. Symp. Ser. **339**, 271 (1987).
86. DAULTON, R.A.C., and R.F. CURTIS: The Effects of *Tagetes* spp. on *Meloidogyne javanica* in Southern Rhodesia. Nematologica **9**, 357 (1963).

87. D'AURIA, M., A. DE MICO, F. D'ONOFRIO, and G. PIANCATELLI: Photochemical Approach to Naturally Occurring Bithiophenes. Synthesis of a Bithienyl Component of *Tagetes erecta*. Gazz. Chim. Ital. **116**, 747 (1986).
88. — — — —: Synthesis of Naturally Occurring Bithiophenes: a Photochemical Approach. J. Org. Chem. **52**, 5243 (1987).
89. — — — —: Photochemical Synthesis of 5-Ethynyl-5′-(1-propynyl)-2,2′-bithiophene. Synth. Commun. **17**, 491 (1987).
90. — — — —: Photochemical Synthesis of Bithienyl derivatives. J. Chem. Soc., Perkin Trans. I, **1987** 1777.
91. DAVIES, G.M., P.S. DAVIES, W.E. PAGET, and J.M. WARDLEWORTH: Borinic Acids: Novel intermediates in Regiospecific Synthesis of Biaryls. Tetrahedron Lett. **1976**, 795.
92. DHAWAN, S.N., J. JAWORSKI, and J. KAGAN: The Singlet Oxygen Sensitizing Ability of Several Butadiynes and Thiophene Derivatives Compared to 8-Methoxypsoralen and Methylene Blue. Photobiophys. Photobiochem. **8**, 25 (1984).
93. DICOSMO, F., R.A. NORTON, and G.H.N. TOWERS: Fungal Culture-Filtrate Elicits Aromatic Polyacetylenes in Plant Tissue Culture. Naturwiss. **69**, 550 (1982).
94. DICOSMO, F., G.H.N. TOWERS, and J. LAM: Photoinduced Fungicidal Activity Elicited by Naturally Occurring Thiophene Derivatives. Pestic. Sci. **13**, 589 (1982).
95. DOMINGUEZ, X.A., G. VAZQUEZ, and R.N. BARUAH: Constituents from *Chrysactinia mexicana*. J. Nat. Prod. **48**, 681 (1985).
96. DOWNUM, K.R.: Personal Communication.
97. DOWNUM, K.R., R.E.W. HANCOCK, and G.H.N. TOWERS: Mode of Action of α-Terthienyl on *Escherichia coli:* Evidence for a Photodynamic Effect on Membranes. Photochem. Photobiol. **36**, 517 (1982).
98. DOWNUM, K.R., R.E.W. HANCOCK, and G.H.N. TOWERS: Photodynamic Action on *Escherichia coli* of Natural Acetylenic Thiophenes, Particularly 5-(buten-1-ynyl)-2,2′-bithienyl. Photobiochem. Photobiophys. **6**, 145 (1983).
99. DOWNUM, K.R., D.J. KEIL, E. RODRIGUEZ: Distribution of Acetylenic Thiophenes in the Pectidinae. Biochem. Syst. Ecol. **13**, 109 (1985).
100. DOWNUM, K.R., D. PROVOST, and L. SWAIN: Acetylenic Thiophenes and C_4 Photosynthesis: Their Evolutionary Relationship in the Asteraceae. Bioact. Mol. (Chem. Biol. Nat.-Occurring Acetylenes Relat. Compd.) **7**, 151 (1988).
101. DOWNUM, K.R., G.A. ROSENTHAL, and G.H.N. TOWERS: Phototoxicity of the Allelochemical, α-terthienyl, to larvae of *Manduca sexta* (L.) (Sphingidae). Pest. Biochem. Physiol. **22**, 104 (1984).
102. DOWNUM, K.R., and G.H.N. TOWERS: Analysis of Thiophenes in the Tageteae (Asteraceae) by HPLC. J. Nat. Prod. **46**, 98 (1983).
103. EVANS, C.H., and J.C. SCAIANO: Photochemical Generation of Radical Cations from α-Terthienyl and Related Thiophenes: Kinetic Behavior and Magnetic Field Effects on Radical-Ion Pairs in Micellar Solution. J. Am. Chem. Soc. **112**, 2694–2701 (1990).
104. EVANS, C., D. WEIR, J.C. SCAIANO, A. MAC EACHERN, J.T. ARNASON, P. MORAND, B. HELLEBONE, L.C. LEITCH, and B.J.R. PHILOGENE. Photochemistry of the Botanical Phototoxin α-Terthienyl and Some Related Compounds. Photochem. Photobiol. **44**, 441 (1986).
105. FLORES, H.E., J.J. PICKARD, and M.W. HOY: Production of Polyacetylenes and Thiophenes in Heterotrophic and Photosynthetic Root Cultures of Asteraceae. Bioact. Mol. (Chem. Biol. Nat.-Occurring Acetylenes Relat. Compd.) **7**, 233 (1988).
106. FOOTE, C.S.: Type I and Type II Mechanisms of Photodynamic Action. ACS Symp. Ser. **339**, 22 (1987).

107. GARCIA, F.J., E. YAMAMOTO, Z. ABRAMOWSKI, K. DOWNUM, and G.H.N. TOWERS: Comparison of the Phototoxicity of α-Terthienyl with that of a Selenium and of an Oxygen Analogue. Photochem. Photobiol. **39**, 521 (1984).

108. GOMMERS, F.J.: Increase of the Nematocidal Activity of α-Terthienyl and Related Compounds by Light. Nematologica **18**, 458 (1972).

109. —: Nematocidal Principles in Compositae. Meded. Landbouwhogesch. Wageningen **73-17**, 1 (1973).

110. GOMMERS, F.J., and J. BAKKER: Mode of Action of α-Terthienyl and Related Compounds May Explain the Suppressant Effects of *Tagetes* species on Populations of Free Living Endoparasitic Plant Nematodes. Bioact. Mol. (Chem. Biol. Nat.-Occurring Acetylenes Relat. Compd.) **7**, 61 (1988).

111. — —: Physiological Diseases Induced by Plant Responses or Products. In: Diseases of Nematodes (G.O. Poinar, Jr. and H.-B. Jansson, eds), Vol 1, p. 1. Boca Raton, FA: CRC Press. (1989).

112. GOMMERS, F.J., J. BAKKER, and L. SMITS: Effects of Singlet Oxygen Generated by the Nematicidal Compound α-Terthienyl from *Tagetes* on the Nematode *Aphelenchus avenae*. Nematologica **26**, 369 (1980).

113. GOMMERS, F.J., J. BAKKER, and H. WYNBERG. Dithiophenes as Singlet Oxygen Sensitizers. Photochem. Photobiol. **35**, 615 (1982).

114. GOMMERS, F.J., and D.J.M. VOOR IN 'T HOLT: Chemotaxonomy of Compositae Related to Their Host Suitability for *Pratylenchus penetrans*. Neth. J. Pl. Path. **82**, 1 (1976).

115. GONG, H.-H., J. KAGAN, R. SEITZ, A.B. STOKES, F.A. MEYER, and R.W. TUVESON: The Phototoxicity of Phenylheptatriyne: Oxygen Dependent Hemolysis of Human Erythrocytes and Inactivation of *Escherichia coli*. Photochem. Photobiol. **47**, 55 (1988).

116. GRAHAM, K., E.A. GRAHAM, and G.H.N. TOWERS. Cercaricidal Activity of Phenylheptatriyne and α-Terthienyl, Naturally Occurring Compounds in Species of Asteraceae (Compositae). Can. J. Zool. **58**, 1955 (1980).

117. GRONEMAN, A.F., M.A. POSTHUMUS, L.G.M.T. TUINSTRA, and W.A. TRAAG: Identification and Determination of Metabolites in Plant Cell Biotechnology by Gas Chromatography and Gas Chromatography/Mass Spectrometry. Application to Nonpolar Products of *Chrysanthemum cinerariaefolium* and *Tagetes* species. Anal. Chim. Acta **163**, 43 (1984).

118. HANDELÉ, M.J.: The Synthesis, Spectra and Nematocidal Activity of 1,2-Dithienylethenes and 1-Phenyl-2-Thienylethenes. Thesis, Wageningen Agricultural University, The Netherlands (1972).

119. HARVEY, Jr., J.: Method for the Control of Plant Growth. **US Pat** 3,086,854, Apr. 23, 1963.

120. HASSPIELER, B.M., J.T. ARNASON, and A.E.R. DOWNE. Toxicity, Localization and Elimination of the Phototoxin, Alpha-Terthienyl, in Mosquito Larvae. J. Am. Mosq. Control Assoc. **4**, 479 (1988).

121. HELSPER, J.P.F.G., D.H. KETEL, A.C. HULST, and H. BRETELER: Production and Secretion of Thiophenes by Differentiated Cell Cultures of *Tagetes*. Bioact. Mol. (Chem. Biol. Nat.-Occurring Acetylenes Relat. Compd.) **7**, 279 (1988).

122. HERZ, W., P. KULANTHAIVEL, and V.L. GOEDKEN: Structures of the Ratibidanolides, Sesquiterpene Lactones With a New Carbon Skeleton, and Unusual Xanthanolides from *Ratibida columnifera*. J. Org. Chem. **50**, 610 (1985).

123. HOGSTAD, S., J. LOEHRE, and T. ANTHONSEN: Possible Confusion of Pyrethrins with Thiophenes in *Tagetes* Species. Acta Chem. Scand., Ser. B **B38**, 902 (1984).

124. HORN, D.H.S., and J.A. LAMBERTON: The Nematocidal Principles of Tagetes Roots. Austral. J. Chem. **16**, 475 (1963).

125. HOTZ, C.Z., P. KOVACIC, I.A. KHOURY: Synthesis and Properties of Polythienylenes. J. Polym. Sci., Polym. Chem. Ed. **21**, 2617 (1983).

126. HUDSON, J.B.: Antiviral Compounds from Plants. Boca Raton, FA: CRC Press (1989).

127. —: Plant Photosensitizers with Antiviral Properties. Antivir. Res. **12**, 55 (1989).

128. HUDSON, J.B., E.A. GRAHAM, G. CHAN, A.J. FINLAYSON, and G.H.N. TOWERS: Comparison of the Antiviral Effects of Naturally Occurring Thiophenes and Polyacetylenes. Planta Medica **6**, 453 (1986).

129. HUDSON, J.B., E.A. GRAHAM, N. MIKI, L. HUDSON, and G.H.N. TOWERS: Antiviral Activity of the Photoactive Thiophene α-Terthienyl. Photochem. Photobiol. **44**, 477 (1986).

130. HUDSON, J.B., E.A. GRAHAM, N. MIKI, G.H.N. TOWERS, L.L. HUDSON, R. ROSSI, A. CARPITA, and D. NERI (1989) Photoactive Antiviral and Cytotoxic Activities of Synthetic Thiophenes and Their Acetylenic Derivatives. Chemosphere **18**, 2317 (1989).

131. HUDSON, J.B., E.A. GRAHAM, and G.H.N. TOWERS: Investigation of the Antiviral Action of the Photoactive Compound Phenylheptatriyne. Photochem. Photobiol. **43**, 27 (1986).

132. HUDSON, J.B., and G.H.N. TOWERS: Antiviral Properties of Photosensitizers. Photochem. Photobiol. **48**, 289 (1988).

133. HUDSON, J.B., G.H.N. TOWERS, Z. ABRAMOWSKI, L. HUDSON, R. ROSSI, A. CARPITA, and D. NERI: Ultraviolet-Mediated Antibiotic Activity of Synthetic Thiophenes and their Acetylenic Derivatives. Chemosphere **18**, 2317 (1989).

134. ITO, Y., T. KONOIKE, and T. SAEGUSA. Synthesis of 1,4-Diketones by the Reaction of Silyl Enol Ethers with Ag_2O. A Regiospecific Formation of Silver (I) Enolate Intermediates. J. Amer. Chem. Soc. **97**, 649 (1975).

135. IYENGAR, S., J.T. ARNASON, B.J.R. PHILOGENE, P. MORAND, N.H. WERSTIUK, and G. TIMMINS: Toxicokinetics of the Phototoxic Allelochemical Alpha-Terthienyl in Three Herbivorous Lepidoptera. Pestic. Biochem. Physiol. **29**, 1 (1987).

136. JAIN, S., and P. SINGH: A Dithienylacetylene Ester from *Eclipta erecta* Linn. Indian J. Chem., Sect. B **27B**, 99 (1988).

137. JAKUPOVIC, J., N. MISRA, T.V. CHAU THI, F. BOHLMANN, and V. CASTRO: Cuauthemone Derivatives from *Tessaria integrifolia* and *Pluchea symphytifolia*. Phytochemistry **24**, 3053 (1985).

138. JAYASURIYA, N.: Polythiophenes: Synthetic Approaches. Diss. Abstr. Int. B. **49**, 3196 (1988).

139. JAYASURIYA, N., and J. KAGAN: The Synthesis of Bithienyls and Terthienyls by Nickel-Catalyzed Coupling of Grignard reagents. Heterocycles **24**, 2261 (1986).

140. — —: The Synthesis of 2,3′:2′,3″-, 2,3′:4′,3″-, 2,3′:5′,3″-, and 2,2′:4′,3″-Terthienyls. Heterocycles **24**, 2901 (1986).

141. JAYASURIYA, N., J. KAGAN, J.E. OWENS, E.P. KORNAK, and D.M. PERRINE. The Photocyclization of Terthiophenes. J. Org. Chem. **54**, 4203 (1989).

142. JENSEN, S. L., and N. A. SÖRENSEN: Studies Related to Naturally Occurring Acetylene Compounds. 29. Preliminary Investigations in the Genus *Bidens*: I. *Bidens radiata* Thuill and *Bidens ferulaefolia* (Jacq.) DC. Acta Chim. Scand. **15**, 1885 (1961).

143. JENTE, R., E. RICHTER, F. BOSOLD, and G.A. OLATUNJI: Experiments on Biosynthesis and Metabolism of Acetylenes and Thiophenes. Bioact. Mol. (Chem. Biol. Nat.-Occurring Acetylenes Relat. Compd.) **7**, 187 (1988).

144. JENTE, R., G.A. OLATUNJI, and F. BOSOLD: Formation of Natural Thiophene Derivatives from Acetylenes by *Tagetes patula*. Phytochemistry **20**, 2169 (1981).

145. KAGAN, J., *et al.* Unpublished Results.

146. Kagan, J., and S.K. Arora: 2,5-Di(2′-thienyl)furan and an Improved Synthesis of Alpha-Terthienyl. Heterocycles **20**, 1941 (1983).

147. — —: The Synthesis of Alpha-Thiophene Oligomers by Oxidative Coupling of 2-Lithiothiophenes. Heterocycles **20**, 1937 (1983).

148. — —: The Synthesis of Alpha-Thiophene Oligomers via Organoboranes. Tetrahedron Letters **24**, 4043 (1983).

149. Kagan, J., S.K. Arora, and A. Ustunol: 2,2′:5′,2″-Terthiophene-5-carboxylic acid and 2,2′:5′,2″-Terthiophene-5,5″-dicarboxylic acid. J. Org. Chem. **48**, 4076 (1983).

150. Kagan, J., M. Bazin, and R. Santus. Photosensitization with α-Terthienyl: Production of Superoxide Ion in Aqueous Media. J. Photochem. Photobiol. B. **3**, 165 (1989).

151. Kagan, J., W.J. Bennett, E.D. Kagan, J.L. Maas, S.A. Sweeney, I.A. Kagan, E. Seigneurie, and V. Bindokas: Alpha-terthienyl as Photoactive Insecticide: Toxic Effects on Non-Target Organisms. ACS Symp. Ser. **339**, 176 (1987).

152. Kagan, J., J.-P. Beny, G. Chan, S.N. Dhawan, J.A. Jaworski, E.D. Kagan, P.D. Kassner, M. Murphy, and J.A. Rogers, The Phototoxicity of Some 1,3-Butadiynes and Related Thiophenes Against Larvae of the Mosquito *Aedes aegypti* and the Fruitfly *Drosophila melanogaster*. Insect Sci. Application **4**, 377 (1983).

153. Kagan, J., and G. Chan: The Photoovicidal Activity of Plant Components Toward *Drosophila melanogaster*. Experientia **39**, 402 (1983).

154. Kagan, J., and R. Gabriel: *Candida utilis* as a Convenient and Safe Substitute for the Pathogenic Yeast *Candida albicans* in Daniels' Phototoxicity Test. Experientia **36**, 587 (1980).

155. Kagan, J., R. Gabriel, and S.A. Reed: Alpha-terthienyl, a Non-Photodynamic Phototoxic Compound. Photochem. Photobiol. **31**, 465 (1980).

156. Kagan, J., M. Hasson and F. Grynspan: The Inactivation of Acetylcholinesterase by Alpha-Terthienyl and Ultraviolet Light. Studies *in vitro* and in Larvae of the Mosquito *Aedes aegypti*. Biochem. Biophys. Acta **802**, 442 (1984).

157. Kagan, J., P.A. Kagan, and H. Buhse, Jr.: Light-Dependent Toxicity of Alpha-Terthienyl and Anthracene Toward Late Embryonic Stages of *Rana pipiens*. J. Chem. Ecol. **10**, 1117 (1984).

158. Kagan, J., E.D. Kagan, and E. Seigneurie: Alpha-Terthienyl, a Powerful Fish Poison with Light-Dependent Activity. Chemosphere **15**, 49 (1986).

159. Kagan, J., E. Kagan, S. Patel, D. Perrine, V. Bindokas: Light-Dependent Effects of Alpha-Terthienyl in Eggs, Larvae, and Pupae of Mosquito *Aedes aegypti*. J. Chem. Ecol. **13**, 593 (1987).

160. Kagan, J., K. Tadema-Wielandt, G. Chan, S.N. Dhawan, J. Jaworski, I. Prakash, and S.K. Arora: Oxygen Requirement for Near-uv Mediated Cytotoxicity of Phenylheptatriyne to *Escherichia coli*. Photochem. Photobiol. **39**, 465 (1984).

161. Kagan, J., and R.W. Tuveson: Are There any Photocytotoxic Reactions of Phenylheptatriyne Which are not Oxygen Dependent?. Bioact. Mol. (Chem. Biol. Nat.-Occurring Acetylenes Relat. Compd.) **7**, 71 (1988).

162. Kato, A.: Secondary Organic-Electrolyte Battery. Jpn. Kokai Tokkyo Koho JP 61110975 A2 29 May 1986.

163. Kellogg, R.M., and H. Wynberg: The Photochemistry of Thiophenes. V. Investigation of Phenylthiophene Photorearrangements by Deuterium Labeling Techniques. J. Amer. Chem. Soc. **89**, 3495 (1967).

164. Kellogg, R.M., and H. Wynberg: Quenching of an Excited Singlet State of 2-Phenylthiophene. Tetrahedron Letters **1968**, 5895.

165. Ketel, D.H.: Accumulation of Thiophenes by Cell Cultures of *Tagetes patula* and the

Release of 5-(4-Hydroxy-1-Butynyl)-2,2'-Bithiophene into the Medium. Planta Med. **54**, 400 (1988).

166. KETEL, D.H., and H. BRETELER: Morphogenesis and Thiophene Production in Cell Cultures of *Tagetes* species. Bioact. Mol. (Chem. Biol. Nat.-Occurring Acetylenes Relat. Compd.) **7**, 267 (1988).

167. KOOREMAN, H.J., and H. WYNBERG. The Chemistry of Polythienyls. Part III. The Synthesis of Terthienyls. Rec. Trav. Chim. **86**, 37 (1967).

168. KRISHNASWAMY, N.R., and S. PRASANNA: Occurrence of Desmethylwedelolactone and 2-Formyl-α-Terthienyl in *Eclipta alba* and the Facile Oxidation of α-Terthienylmethanol. Indian J. Chem. **8**, 761 (1970).

169. KRISHNASWAMY, N.R., T.R. SESHADRI, B.R. SHARMA: The Structure of a New Polythienyl from *Eclipta alba.* Tetrahedron Lett. **1966**, 4227.

170. LABLACHE-COMBIER, A.: Photochemical Reactions of Thiophenes. In: Thiophene and its Derivatives, Part 1 (S. Gronowitz, ed.), p. 745. New York: John Wiley (1985).

171. LAM, J., and T. THOMASEN: Complexing Agents for Protection of Highly Conjugated Compounds Against Photodegradation. Bioact. Mol. (Chem. Biol. Nat.-Occurring Acetylenes Relat. Compd.) **7**, 47 (1988).

172. LIN, J.W.P., and L.P. DUDEK: Synthesis and Properties of Poly (2,5-Thienylene). J. Polym. Sci., Polym. Chem. Ed. **18**, 2869 (1980).

173. MACRAE, W.D., G.F.Q. CHAN, C.-K. WAT, G.H.N. TOWERS, and J. LAM: Examination of Naturally Occurring Polyacetylenes and α-Terthienyl for Their Ability to Induce Cytogenetic Damage. Experientia **36**, 1096 (1980).

174. MACRAE, W.D., D.A.J. IRWIN, T. BISALPUTRA, and G. H. N. TOWERS: Membrane Lesions in Human Erythrocytes Induced by the Naturally Occurring Compounds α-Terthienyl and Phenylheptatriyne. Photobiochem. Photobiophys. **1**, 309–318 (1980).

175. MAKJANIC, J., R.D. VIS, A.F. GRONEMAN, F.J. GOMMERS, and S. HENSTRA: Investigation of P and S Distribution in the Roots of *Tagetes patula* L. Using Micro-PIXE. J. Exp. Bot. **39**, 1523 (1988).

176. MALDONADO, Z., M. HOENEISEN, and M. SILVA: A Dithiophene from *Aphyllocladus denticulauts.* Phytochemistry **27**, 2993 (1988).

177. MANDOLI, D.F. and W.R. BRIGGS, Fiber-Optic Plant Tissues: Spectral Dependence in Dark-Grown and Green Tissues. *Photochem. Photobiol.* **39**, 709–715 (1984).

178. MARCHANT, Y.Y., and G.K. COOPER: Structure and Function Relationships in Polyacetylene Photoactivity. ACS Symp. Ser. **339**, 241 (1987).

179. MARCHANT, Y.Y., and G.H.N. TOWERS: Phototoxicity of Polyacetylenes to *Cryptococcus Laurentii.* Biochem. Syst. Ecol. **14**, 565 (1986).

180. — —: Phylloplane Fungi of Hawaiian Plants and their Photosensitivity to Polyacetylenes from *Bidens* Species. Biochem. Syst. Ecol. **15**, 9 (1987).

181. MCDOUGALL, C., B.J.R. PHILOGENE, J.T. ARNASON, and N. DONSKOV. Comparative Effects of Two Plants Secondary Metabolites on Host-Parasitoid Association. J. Chem. Ecol. **14**, 1239 (1988).

182. MCLACHLAN, D., J.T. ARNASON, B.R. HOLLEBONE, and J. LAM: Excited States of Phototoxic Polyacetylenes Elucidated by Magnetic Circular Dichroism. Photobiochem. Photobiophys. **9**, 233 (1985).

183. MCLACHLAN, D., T. ARNASON, and J. LAM: The Role of Oxygen in Photosensitizations with Polyacetylenes and Thiophene Derivatives. Photochem. Photobiol. **39**, 177 (1984).

184. — — —: Structure-Function Relationships in the Phototoxicity of Acetylenes from the Asteraceae. Biochem. Syst. Ecol. **14**, 17 (1986).

185. McRae, D.G., E. Yamamoto, and G.H.N. Towers: The Mode of Action of Polyacetylene and Thiophene Photosensitizers on Liposome Permeability to Glucose. Biochim. Biophys. Acta **821**, 488 (1985).

186. Metschulat, G., and Sütfeld, R.: Acetyl-CoA: 4-Hydroxybutynylbithiophene O-Acetyltransferase from *Tagetes patula* Seedlings. Z. Naturforsch. **42**, 885 (1987).

187. Moriarty, R.M., O. Prakash, and M. Duncan: Synthesis of 3,2′:5′,3″-Terthiophene and 2,5-Di(3′-Thienyl)furan. Synth. Commun. **15**, 789 (1985).

188. Munoz, L.C., C.O. Castro, C.R. Lopez, A.R. Arias, F. Pignani, and J. Calzada: Potential Natural Nematocides from Plants of the Genus *Tagetes* (Compositae). Ing. Cienc. Quim. **6**, 158 (1982).

189. Murase, M., A. Usuki, Y. Kitahara: Polymeric Photoelectric Transducer. Jpn. Kokai Tokkyo Koho JP 61125090 A2, 12 Jun 1986.

190. Nakayama, J., Y. Nakamura, T. Tajiri, and M. Hoshino: Preparation of Naturally Occurring α-Terthiophenes (2,2:5′,2″-Terthiophenes). Heterocycles **24**, 637 (1986).

191. Norton, R.A., A.J. Finlayson, and G.H.N. Towers: Two Dithiacyclohexadiene Polyacetylenes form *Chaenactis douglasii* and *Eriophyllum lanatum*. Phytochemistry **24**, 356 (1985).

192. — — —: Thiophene Production by Crown Gall and Callus Tissue of *Tagetes patula*. Phytochemistry **24**, 719 (1985).

193. Obata, S., M. Youshikura, T. Washino: Components of the Roots of *Arctium lappa*. Nippon Nogei Kagaku Kaishi **44**, 437–46 (1970).

194. Okuhara, K.: Introduction and Extension of Ethynyl Group using 1,1-dichloro-2,2-Difluoroethylene. A Convenient Route to Lithium Acetylides and Derived Acetylenic Compounds. J. Org. Chem. **41**, 1487 (1976).

195. Onishi, T., S. Yokoishi, and Y. Nonobe: Polythienylene Electrochromic Display Apparatus. Jpn. Kokai Tokkyo Koho JP 62288037 A2 14 Dec 1987.

196. Palaszek, M.: Biosynthesis of Polythienyls in *Tagetes erecta*. Diss. Abstr. B **27**, 727 (1966).

197. Parodi, F.J., N.H. Fischer, and H.E. Flores: Benzofuran and Bithiophenes from Root Cultures of *Tagetes patula*. J. Nat. Prod. **51**, 594 (1988).

198. Patrick, T.B., J.M. Disher, and W.J. Probst: Synthesis and Metalation of 2-Ethynylthiophene. J. Org. Chem. **37**, 4467 (1972).

199. Patrick, T.B., and J.L. Honegger: Synthesis of 5-Ethynyl-2,2′-Bithienyl and Related Compounds. J. Org. Chem. **39**, 3791 (1974).

200. Pensl, R. and Sütfeld, R.: Occurrence of 3,4-Diacetoxybutynylbithiophene in *Tagetes patula* and its Enzymatic Conversion. Z. Naturforsch. **40c**, 3 (1985).

201. Philogene, B.J.R., J.T. Arnason, C.W. Berg, F. Duval, D. Champagne, R.G. Taylor, L.C. Leitch, and P. Morand: Synthesis and Evaluation of the Naturally Occurring Phototoxin, α-Terthienyl, as a Control Agent for Larvae of *Aedes intrudens, Aedes atropalpus* (Diptera: Culicidae) and *Simulium verecundum* (Diptera: Simuliidae). J. Econ. Entomol. **78**, 121 (1985).

202. Philogene, B.J.R., J.T. Arnason, C. W. Berg, F. Duval, and P. Morand: Efficacy of the Plant Phototoxin α-Terthienyl Against *Aedes intrudens* and Effects on Non-Target Organisms. J. Chem. Ecol. **12**, 893 (1986).

203. Rampone, W.M., J.L. McCullough, G.D. Weinstein, G.H.N. Towers, M.W. Berns, and N. Abeysekera: Characterization of Cutaneous Phototoxicity Induced by Topical Alpha-terthienyl and Ultraviolet A Radiation. J. Invest. Dermatol. **87**, 354 (1986).

204. REYFTMANN, J.P., J. KAGAN, R. SANTUS, and P. MORLIERE. Excited State Properties of α-Terthienyl and Related Molecules. Photochem. Photobiol. **41**, 1 (1985).

205. ROSSI, R.: Highly Selective Syntheses of Naturally-Occurring Acetylenes and Their Structural Analogues by Palladium-Catalyzed Carbon-Carbon Bond Forming Reactions. Bioact. Mol. (Chem. Biol. Nat.-Occurring Acetylenes Relat. Compd.) **7**, 29 (1988).

206. ROSSI, R., CARPITA, A., and A. LEZZI: Palladium-Catalyzed Synthesis of Naturally-Occurring Acetylenic Thiophenes and Related Compounds. Tetrahedron **40**, 2773 (1984).

207. RUDISII, D.E., and J.K. STILLE: Palladium-Catalyzed Synthesis of 2-Substituted Indoles. J. Org. Chem. **54**, 5856 (1989).

208. SCAIANO, J.C.: Personal Communication.

209. SCAIANO, J.C., C. EVANS, and J.T. ARNASON: Characterization of the α-Terthienyl Radical Cation: Evidence Against Electron Transfer to Oxygen *in vitro*. J. Photochem. Photobiol., B: Biol. **3**, 411 (1989).

210. SCHEEREN, J.W., P.H.J. OOMS, and R.J.F. NIVARD: A General Procedure for the Conversion of a Carbonyl Group Into a Thione Group with Tetraphosphorus Decasulfide. Synthesis **1973**, 149.

211. SCAIANO, J.C., R.W. REDMOND, B. MEHTA, and J.T. ARNASON: Efficiency of the Photoprocesses Leading to Singlet Oxygen ($^1\Delta_g$) Generation by α-Terthienyl: Optical Absorption, Optoacoustic Calorimetry and Infrared Luminescence Studies. Photochem. Photobiol. **54**, 655 (1990).

212. SHOWA DENKO K.K., Japan. Crystalline Poly(2,5-thienylene) and its Preparation. Jpn. Kokai Tokkyo Koho JP 59221330 A2, 12 Dec 1984.

213. SCHUETZ, R.D., T.B. WAGGONER, and R.U. BYERRUM: Biosynthesis of 2,2′;5′,2″-Terthienyl in the Common Marigold. Biochem. **4**, 436 (1965).

214. SCHULTE, K.E., and S. FOERSTER: Is Bithienylbutynene the Biogenetic Precursor of α-Terthienyl? Tetrahedron Lett. **1966**, 773.

215. SCHULTE, K.E., G. HENKE, G. RÜCKER and S. FOERSTER: Beitrag zur Biogenese des α-Terthienyls. Tetrahedron **24**, 1899 (1968).

216. SCHULTE, K.E., J. REISCH, and L. HÖRNER: Thiophene aus Alkinen, I. Chem. Ber. **95**, 1943 (1962).

217. SELVA, A., A. ARNONE, R. MONDELLI, V. SPRIO, L. CERAULO, S. PETRUSO, S. PLESCIA, and L. LAMARTINA: Cardopatine and Isocardopatine, Two Novel Cyclobutane substances from *Cardopatium corymbosum*. Phytochemistry **17**, 2097 (1978).

218. SINCLAIR, J.A., and H.C. BROWN: Synthesis of Unsymmetrical Conjugated Diynes via the Reaction of Lithium Dialkynyldialkylborates with Iodine. J. Org. Chem. **41**, 1078 (1976).

219. SINCLAIR, J., and T. ARNASON: The effect of Alpha-Terthienyl on Photosynthesis. Can. J. Bot. **60**, 2565 (1982).

220. SINGH, P.: Naturally-occurring Thiophene Derivatives from *Eclipta* Species. Bioact. Mol. (Chem. Biol. Nat.-Occurring Acetylenes Relat. Compd.) **7**, 179 (1988).

221. SINGH, P., A.K. SHARMA, K.C. JOSHI, and F. BOHLMANN: A Further Dithienylacetylene from *Eclipta erecta*. Phytochemistry **24**, 615 (1985).

222. SINGH, S.P., P. SHARMA and L.K. VATS: Light-Dependent Toxicity of the Extract of Plant *Tagetes erecta* and Alpha-Terthienyl Toward Larvae of Mosquito *Culex tritaeniorhynchus*. Toxic. Environ. Chem. **16**, 81 (1987).

223. SKATTEBÖL, L.: Studies Related to Naturally Occurring Acetylene Compounds, XXVI. The synthesis of 5-(1-Propynyl)-2-Formylthiophene, Junipal, and *trans* Methyl 5-(1-Propynyl)-2-Thienylacrylate. Acta Chem. Scand **13**, 1460 (1959).

224. SPRIO, V., S. PLESCIA, and S. PETRUSO: Chemical Investigation of the Roots of *Cardopatium corymbosum*. Ann. Chim. (Rome) **62**, 568 (1972).
225. STETTER, H.: Catalyzed Addition of Aldehydes to Activated Double Bonds — A New Synthetic Approach. Angew. Chem., Int. Ed. Engl. **15**, 639 (1976).
226. STICH, H.F., P. LAM, L.W. LO, D.J. KOROPATNICK, and R.H.C. SAN: The Search for Relevant Short Term Bioassays for Chemical Carcinogens:The Tribulation of a Modern Sisyphus. Can. J. Genet. Cytol. **17**, 471 (1975).
227. SÜTFELD, R. and G.H.N. TOWERS: 5-(4-Acetoxy-1-Butynyl)-2,2′-Bithiophene:Acetate Esterase from *Tagetes patula*. Phytochemistry **21**, 277 (1982).
228. SÜTFELD, R.: Distribution of Thiophene Derivatives in Different Organs of *Tagetes patula* Seedlings Grown under Various Conditions. Planta **156**, 536 (1987).
229. SÜTFELD, R.: Enzymological Investigations into the Metabolism of Bithiophene Derivatives. Bioact. Mol. (Chem. Biol. Nat.-Occurring Acetylenes Relat. Compd.) **7**, 201 (1988).
230. SÜTFELD, R., and H. BRETELER, Effects of Plant Material and Extract Treatment on the Yield of Natural Products from *Tagetes*. Bioact. Mol. (Chem. Biol. Nat.-Occurring Acetylenes Relat. Compd.) **7**, 101 (1988).
231. TAMARU, Y., Y. YAMADA, and Z. YOSHIDA. The Palladium Catalyzed Thienylation of Allylic Alcohols with 2- and 3-Bromothiophenes and their Derivatives. Tetrahedron **35**, 329 (1979).
232. TANG, C.S., C.K. WAT, and G.H.N. TOWERS: Thiophenes and Benzofurans in the Undisturbed Rhizosphere of *Tagetes patula* L. Plant Soil **98**, 93 (1987).
233. TOSI, B., G. LODI, F. DONDI, and A. BRUNI: Thiophene Distribution during the Ontogenesis of *Tagetes patula*. Bioact. Mol. (Chem. Biol. Nat.-Occurring Acetylenes Relat. Compd.) **7**, 209 (1988).
234. TOWERS, G.H.N., T. ARNASON, C.-K. WAT, E.A. GRAHAM, J. LAM, and J.C. MITCHELL: Phototoxic Polyacetylenes and Their Thiophene Derivatives [Effects on human skin]. Contact Dermatitis **5**, 140 (1979).
235. TOWERS, G.H.N., J.T. ARNASON, C.K. WAT, and J.D.H. LAMBERT: Controlling Pests Using a Naturally Occurring Polyacetylene. Can. CA 1173743 A1, 4 Sept 1984; Chem. Abstr. **102**, P41602r (1985).
236. TOWERS, G.H.H., J.T. ARNASON, C.K. WAT, and J.D. LAMBERT: Cercaricidal Compositions Containing a Naturally Occurring Conjugated Polyacetylene and Method for Controlling Cercariae Using It. Can. CA 1169767 A1, 26 Jun 1984; Chem. Abstr. **101**, P146144a (1984).
237. TOWERS, G.H.N., and J.B. HUDSON: Potentially Useful Antimicrobial and Antiviral Phototoxins From Plants. Photochem. Photobiol. **46**, 61 (1987).
238. TOWERS, G.H.N., C.-K. WAT, E.A. GRAHAM, R.J. BANDONI, G.F.Q. CHAN, J.C. MITCHELL and J. LAM: Ultraviolet-Mediated Antibiotic Activity of Species of Compositae Caused by Polyacetylenic Compounds. Lloydia **40**, 487 (1977).
239. TUVESON, R.W., M.R. BERENBAUM, and E.E. HEININGER: Inactivation and Mutagenesis by Phototoxins using *Escherichia coli* Strains Differing in Sensitivity to Near and Far-Ultraviolet Light. J. Chem. Ecol. **12**, 933 (1986).
240. TUVESON, R.W., R.A. LARSON, and J. KAGAN: The Role of Cloned Carotenoid Genes Expressed in *Escherichia coli* in Protecting Against Inactivation by Far-UV, Near-UV, and Specific Phototoxic Molecules. J. Bacteriol. **170**, 4675 (1988).
241. TUVESON, R.W., and G.-R. WANG: Unpublished Results.
242. TYLER, J.: Proceedings of the Root-Knot Nematode Conference Held in Atlanta, GA. Pl. Dis. Rep. **109**, 133 (1938).
243. UHLENBROEK, J.H., and J.D. BIJLOO: Investigation on Nematicides I. Isolation and

Structure of a Nematicidal Principle occurring in *Tagetes* roots. Rec. Trav. Chim. **77**, 1004 (1958).

244. UHLENBROEK, J.H., and J.D. BIJLOO: Investigation on Nematicides II. Structure of a Second Nematicidal Principle Isolated from *Tagetes* Roots. Rec. Trav. Chim. **78**, 382 (1959).

245. — —: Investigation on Nematicides III. Polythienyls and Related Compounds. Rec. Trav. Chim. **79**, 1181 (1960).

246. VAN BOLHUIS, F., H. WYNBERG, E.E. HAVINGA, E.W. MEIJER, and E.G.J. STARING: The X-Ray Structure and MNDO Calculations of α-Terthienyl: a Model for Polythiophenes. Synth. Met. **30**, 381 (1989).

247. VAN FLEET, D.S.: Histochemistry of Plants in Health and Disease: In: Structural and Functional Aspects of Phytochemistry (V.C. Runeckles and T.C. Tso, eds), p. 165. New York: Academic Press (1972).

248. WALTMAN, R.J., J. BARGON, and A.F. DIAZ: Electrochemical Studies of Some Conducting Polythiophene Films. J. Phys. Chem. **87**, 1459 (1983).

249. WANG, T.P., and J. KAGAN: Ageing of Human Erythrocytes: Effects on Photosensitized Hemolysis. Chemosphere **19**, 1345 (1990).

250. WANG, T.P., J. KAGAN, R.W. TUVESON, and G.R. WANG: α-Terthienyl Photosensitizes Damage to pBR 322 DNA. Photochem. Photobiol., in press (1990).

251. WARREN, R.A., J.B. HUDSON, K.R. DOWNUM, E.A. GRAHAM, R. NORTON, and G.H.N. TOWERS: Bacteriophages as Indicators of the Mèchanism of Action of Photosensitizing Agents. Photobiochem. Photobiophys. **1**, 385 (1980).

252. WASHINO, T., M. YOSHIKURA, and S. OBATA: New Sulfur-containing Acetylenic compounds. Agric. Biol. Chem. **50**, 263 (1986).

253. WASHINO, T., M. YOSHIKURA, and S. OBATA: Synthesis of 5′-(1-Propynyl)-2,2′-Bithienyl-5-yl Derivatives. Agr. Biol. Chem. **50**, 565 (1986).

254. WAT, C.-K., R.K. BISWAS, E.A. GRAHAM, L. BOHM, G.H.N. TOWERS, and E.R. WAYGOOD. J. Natl. Prod. **42**, 103 (1977).

255. WAT, C.-K., W.D. MACRAE, E. YAMAMOTO, G.H.N. TOWERS, and J. LAM: Phototoxic Effects of Naturally Occurring Polyacetylenes and α-Terthienyl on Human Erythrocytes. Photochem. Photobiol. **32**, 167 (1980).

256. WAT, C.K., S.K. PRASAD, E.A. GRAHAM, S. PARTINGTON, T., G.H.N. TOWERS, J. LAM: Photosensitization of Invertebrates by Natural Polyacetylenes. Biochem. Syst. Ecol. **9**, 59 (1981).

257. WYNBERG, H., R.M. KELLOGG, H. VAN DRIEL, and G.E. BEEKHUIS: The Photochemistry of Thiophenes. VI. Photorearrangement of Phenylmethylthiophenes. J. Amer. Chem. Soc. **89**, 3498 (1967).

258. — — — —: The Photochemistry of Thiophenes. III. Photochemical Rearrangement of Arylthiophenes. J. Amer. Chem. Soc. **88**, 5047 (1966).

259. — — — —: The Photochemistry of Thiophenes. VII. Observations on the Mechanism of Arylthiophene Rearrangements. J. Amer. Chem. Soc. **89**, 3501 (1967).

260. WYNBERG, H., and J. METSELAAR: A Convenient Route to Polythiophenes. Synth. Commun. **14**, 1 (1984).

261. WYNBERG, H., and H. VAN DRIEL: The Photochemical Rearrangement of Arylthiophenes. J. Amer. Chem. Soc. **87**, 3998 (1965).

262. — —: Further Evidence for Scrambling in the Photochemical Rearrangement of 2-Phenylthiophenes. Chem. Commun. **1966**, 204.

263. WYNBERG, H., H. VAN DRIEL, R.M. KELLOGG, and J. BUTER: The Photochemistry of Thiophenes. IV. Observations on the Scope of Arylthiophene Rearrangements. J. Amer. Chem. Soc. **89**, 3487 (1967).

264. YAMAMOTO, E., W.D. MACRAE, F.J. GARCIA, and G.H.N. TOWERS: Photodynamic Hemolysis Caused by α-Terthienyl. Planta Med. **50**, 124 (1984).

265. YAMAMOTO, T, K. OSAKADA, T. WAKABAYASHI, and A. YAMAMOTO: Nickel and Palladium Catalyzed Dehalogenating Polycondensation of Dihaloaromatic Compounds with Zinc. A New Route to Poly(2,5-thienylene) and Poly(1,4-phenylene). Makromol. Chem., Rapid Commun. **6**, 671 (1985).

266. YAMAMOTO, T., K. SANECHIKA, and A. YAMAMOTO: Preparation of Thermostable and Electric-Conducting Poly(2,5-thienylene). J. Polym. Sci., Polym. Lett. Ed. **18**, 9 (1980).

267. YAMAMOTO, E., C.-K. WAT, W.D. MACRAE, G.H.N. TOWERS, and C.F.Q. CHAN: Photoinactivation of Human Erthrocyte Enzymes by α-Terthienyl and Phenylheptatriyne, Naturally Occurring Compounds in the Asteraceae. FEBS Letters **107**, 134 (1979).

268. YONEYAMA, H., K. KAWAI, and S. KUWABATA: Light-Localized Deposition of Electroconductive Polymers on n-Type Silicon by Utilizing Semiconductor Photocatalysis. J. Electrochem. Soc. **135**, 1699 (1988).

269. YUMOTO, Y., and S. YOSHIMURA: Synthesis and Electrical Properties of a New Conducting Polythiophene Prepared by Electrochemical Polymerization of α-Terthienyl. Synth. Met. **13**, 185 (1986).

270. ZECHMEISTER, L. and J.W. SEASE: A Blue-Fluorescing Compound, Terthienyl, Isolated from Marigolds. J. Amer. Chem. Soc. **69**, 273 (1947).

(*Received May 21, 1990*)

Author Index

Page numbers printed in *italics* refer to References

Subject Index

Volume 51:

1987. VII, 317 pages. Cloth DM 280,–, öS 1960,–.
ISBN 3-211-81972-X

Contents: M. Gill and W. Steglich: Pigments of Fungi (Macromycetes).

Volume 50:

1986. 71 figures. IX, 261 pages. Cloth DM 210,–, öS 1470,–.
ISBN 3-211-81969-X

Contents: L. Jaenicke and F.-J. Marner: The Irones and Their Precursors – M. Lounasmaa and P. Somersalo: The Condylocarpine Group of Indole Alkaloids – U. Séquin: The Antibiotics of the Pluramycin Group (4*H*-Anthra [1,2-*b*]pyran Antibiotics) – R. M. Wenger: Cyclosporine and Analogues – Isolation and Synthesis – Mechanism of Action and Structural Requirements for Pharmacological Activity – H. Inouye and S. Uesato: Biosynthesis of Iridoids and Secoiridoids.

Volume 49:

1986. VIII, 400 pages. Cloth DM 290,–, öS 2030,–.
ISBN 3-211-81910-X

Contents: R. A. Hill: Naturally Occurring Isocoumarins – R. Wijnsma and R. Verpoorte: Anthraquinones in the Rubiaceae – H. Chr. Krebs: Recent Developments in the Field of Marine Natural Products with Emphasis on Biologically Active Compounds.

All Volumes and Cumulative Index 1–20 available

Price reduction for subscribers: 10%

Special reduced price (20% reduction) for the complete Series Vols. 1–56 incl. the Cumulative Index to Vols. 1–20

Springer-Verlag Wien New York

Mölkerbastei 5, A-1011 Wien
175 Fifth Avenue, New York, NY 10010, U.S.A.
Heidelberger Platz 3, D-1000 Berlin 33
37-3, Hongo 3-chome, Bunkyo-ku, Tokyo 113, Japan

Applied Magnetic Resonance

Published by the
Department of General Physics and Astronomy
& Kazan Physical-Technical Institute
of the Academy of Sciences of USSR

ISSN 0937-9347
Title No. 723

Editor: **Kev M. Salikhov,**
Zavoisky Physical-Technical Institute
of Academy of Sciences of the USSR
Sibirsky tract, 1017, Kazan, 420029, USSR

Associate Editors:
U. Haeberlen,
Max-Planck-Institut für Medizinische Forschung
Jahnstrasse 29, D-6900 Heidelberg, Germany
Keith R. Carduner,
Ford Motor Company, Scientific Research Labs.,
P.O.Box 2052, Dearborn, MI 48121-2063, USA

and an international Editorial Board

Applied Magnetic Resonance provides an international forum for the application of magnetic resonance in physics, chemistry, biology, medicine, geochemistry, ecology, engineering, and related fields.
AMR publishes original articles with a strong emphasis on new applications of the technique and on new experimental methods. Routine applications to structural chemistry are outside the scope of the journal.
Review articles are invited on methods and applications of NMR, NQR, EPR, Moessbauer spectroscopy, etc. Special issues on selected topics are published under the guidance of guest editors.
A section is devoted to book reviews and Letters to the Editor. Information on conferences and new technical equipment are also accepted.
AMR is the first international scientific journal in English produced and published in the USSR. It is distributed outside the USSR by Springer-Verlag Wien New York. The editors and Springer-Verlag hope that this journal is only the first step in bridging the gap between scientists in the East and the West.

Subscription Information:
1991. Vol. 2 (4 issues): DM 440,–, öS 3080,–
plus carriage charges
Prices are subject to change without notice

Distributed by Springer-Verlag Wien New York